An OPUS book

The Life and Times of Liberal Democracy

C. B. MACPHERSON

The Life and Times of Liberal Democracy

Oxford New York
OXFORD UNIVERSITY PRESS

Oxford University Press, Walton Street, Oxford OX2 6DP
Oxford New York Toronto
Delhi Bombay Calcutta Madras Karachi
Kuala Lumpur Singapore Hong Kong Tokyo
Nairobi Dar es Salaam Cape Town
Melbourne Auckland
and associated companies in
Beirut Berlin Ibadan Nicosia

Oxford is a trade mark of Oxford University Press

First published 1977 as an Oxford University Press paperback
and simultaneously in a hardback edition
Paperback reprinted 1979, 1980, 1984, 1986

British Library Cataloguing in Publication Data
Macpherson, Crawford Brough
The life and times of liberal democracy.
1. Democracy
I. Title
321.8 JC423 77–30093
ISBN 0–19–289106–5

Printed in Great Britain by
The Guernsey Press Co. Ltd.
Guernsey, Channel Islands

Preface

Readers may wonder at the shortness of this book. 'The Life and Times', in a title, usually signals a book ten times as long as this one. But no such length is required by my design, which is to set out in bold relief the essence of liberal democracy as it now is conceived, and as it has been and may be conceived. For this purpose brevity is better than exhaustive detail. I hope however that my analysis is substantial enough both to establish the patterns I have found and to justify the criticism and praise from which I have seen no reason to abstain.

Successive preliminary versions of this work have been presented for criticism in several universities: the earliest, most tentative, version at the University of British Columbia, and subsequent versions, each profiting from earlier criticisms, at the Institute of Advanced Studies of the Australian National University, the Institute of Philosophy of Aarhus University, and the University of Toronto. Parts of it have also been presented and effectively criticized at several United States universities and some other Canadian universities. Colleagues and students who took part in the discussions in all those countries will recognize how much I have benefited from their criticisms. Some will wish I had benefited more. But I thank them all.

University of Toronto
4 October 1976

C.B.M.

Contents

I

Models and Precursors

THE NATURE OF THE INQUIRY.

It is not usual to embark on a 'Life and Times' until the subject's life is over. Is liberal democracy, then, to be considered so nearly finished that one may presume now to sketch its life and times? The short answer, prejudging the case I shall be putting, is: 'Yes', if liberal democracy is taken to mean, as it still very generally is, the democracy of a capitalist market society (no matter how modified that society appears to be by the rise of the welfare state); but 'Not necessarily' if liberal democracy is taken to mean, as John Stuart Mill and the ethical liberal-democrats who followed him in the late nineteenth and early twentieth centuries took it to mean, a society striving to ensure that all its members are equally free to realize their capabilities. Unfortunately, liberal democracy can mean either. For 'liberal' can mean freedom of the stronger to do down the weaker by following market rules; or it can mean equal effective freedom of all to use and develop their capacities. The latter freedom is inconsistent with the former.

The difficulty is that liberal democracy during most of its life so far (a life which, I shall argue, began only about a hundred and fifty years ago even as a concept, and later as an actual institution) has tried to combine the two meanings. Its life began in capitalist market societies, and from the beginning it accepted their basic unconscious assumption, which might be paraphrased 'Market maketh man'. Yet quite early on, as early as John Stuart Mill in the mid-nineteenth century, it pressed the claim of equal individual rights to self-development, and justified itself largely by that claim. The two ideas of liberal democracy have since then been held together uneasily, each with its ups and downs.

So far, the market view has prevailed: 'liberal' has consciously or unconsciously been assumed to mean 'capitalist'. This is true even though ethical liberals, from Mill on, tried to combine market freedom with self-developmental freedom, and tried to subordinate the former to the latter. They failed, for reasons explored in Chapter III.

Here I am simply suggesting that a liberal position need not be taken to depend forever on an acceptance of capitalist assumptions, although historically it has been so taken. The fact that liberal values grew up in capitalist market societies is not in itself a reason why the central ethical principle of liberalism—the freedom of the individual to realize his or her human capacities—need always be confined to such societies. On the contrary, it may be argued that the ethical principle, or, if you prefer, the appetite for individual freedom, has outgrown its capitalist market envelope and can now live as well or better without it, just as man's productive powers, which grew so enormously with competitive capitalism, are not lost when capitalism abandons free competition or is replaced by some form of socialism.

I shall suggest that the continuance of anything that can properly be called liberal democracy depends on a downgrading of the market assumptions and an upgrading of the equal right to self-development. I think there is some prospect of this happening. But it is far from certain that it will happen. So I have felt justified in keeping the sombre title 'Life and Times'.

My main concern in this short work is to examine the limits and possibilities of liberal democracy. Let me explain now why I have done this in terms of models, and why I have chosen certain models as appropriate and sufficient. This will lead into a consideration of certain earlier models which I have relegated to the position of precursors of liberal democracy.

THE USE OF MODELS

(i) *Why models?*

I am using the term 'model' in a broad sense, to mean a

theoretical construction intended to exhibit and explain the real relations, underlying the appearances, between or within the phenomena under study. In the natural sciences, which are mostly concerned with phenomena not variable by human will or by social change, successive models (as those of Ptolemy, Copernicus, Newton, Einstein) are successively fuller and more sufficient explanations of the real, invariant relations. In the social sciences, concerned with phenomena which, within historically shifting limits, are variable by human will, models (or theories, as we may equally well call them) may have two additional dimensions.

First, they may be concerned to explain not only the underlying reality of the prevailing or past relations between wilful and historically influenced human beings, but also the probability or possibility of future changes in those relations. By sorting out main lines of change, and apparently unchanging characteristics, of man and society up to the present, they may try to discern forces of change, and limits of change, which may be expected to operate in the future. Not all the theorists who have formulated laws of change have seen them as operating in a straight line: Machiavelli, for instance, thought in terms of a cyclical movement as the historical pattern of social and political change which could be expected to prevail indefinitely into the future. But ever since the eighteenth-century Enlightenment, with its idea of progress, it has been more usual to think in terms of a straight line. Of the theorists who have seen a single main line of past change, not all have projected it far, if at all, into the future: for instance, such eighteenth-century writers as Montesquieu, Turgot, Millar, Ferguson, and Adam Smith, who glimpsed or formulated the law of four stages of society—hunting, pastoral, agricultural, commercial—were apt to assume that the commercial was the final stage. But in the nineteenth century others, as different as Comte and Marx and Mill, have, with greater or less stringency, projected a main line of past development into the future. Any of these kinds of theory do of course rely explicitly or implicitly on models.

The second additional dimension of models in political theorizing is an ethical one, a concern for what is desirable or good or right. The outstanding models in political science, at least from Hobbes on, have been both explanatory and justificatory or advocatory. They are, in different proportions, statements about what a political system or a political society is, how it does work or could work, and statements of why it is a good thing, or why it would be a good thing to have it or to have more of it. Some democratic theorists have seen clearly enough that their theories are such a mixture. Some have not, or have even denied it. Those who start from the tacit assumption that whatever is, is right, are apt to deny that they are making any value judgement. Those who start from the tacit assumption that whatever is, is wrong, give great weight to their ethical case (while trying to show that it is practicable). And between the two extremes there is room for a considerable range of emphasis.

In any case, to show that a model of a political system or a society, whether the existing one or one not now existing but desired, is practicable, that is, that it can be expected to work well over a fairly long run, one must make some assumptions about the human beings by whom and with whom it is going to run. What kind of political behaviour are they capable of? This is obviously a crucial question. A political system that demanded, for instance, that the citizens have more rationality or more political zeal than they now demonstrably have, and *more than they could be expected to have in any attainable social circumstances*, would not be worth much advocacy. The stipulation I have just emphasized is important. We are not necessarily limited to the way people behave politically now. We are not limited to that if we can show reasons for expecting that that could change with changes in, for instance, the technological possibilities and the economic relations of their society.

Most, though not all, political theorists of all persuasions—conservative traditionalists, liberal individualists, radical reformists, and revolutionaries—have understood very well that the workability of any political system depends largely on how all the other institutions, social and economic, have shaped, or might shape, the people with whom and by whom the political system must operate. On this, writers as different as Burke and Mill and Marx are in agreement, although most of the earlier liberal theorists, from say Locke to Bentham, paid little attention to this. And it has generally been seen, at least in the nineteenth and twentieth centuries, that the most important way in which the whole bundle of social institutions and social relations shapes people as political actors is in the way they shape people's consciousness of themselves. For instance, when, as in the Middle Ages and for some time after, the prevailing social arrangements have induced virtually everyone to accept an image of the human being as human by virtue of his accepting the obligations of his rank or his 'station in life', a traditional hierarchical political system will work. When a commercial and an industrial revolution have so altered things that that image is no longer accepted, a different image is required. If it is an image of man as essentially a maximizing consumer and appropriator we get a new consciousness, which permits and requires a quite different political system. If, later, in revulsion against the results of this, people come to think of themselves in some other way, some other political system becomes possible and even needed.

So, in looking at models of democracy—past, present, and prospective—we should keep a sharp look-out for two things: their assumptions about the whole society in which the democratic political system is to operate, and their assumptions about the essential nature of the people who are to make the system work (which of course, for a democratic system, means the people in general, not just a ruling or leading class).

To speak, as I have just done, of 'the society in which a democratic political system is to operate' may seem to suggest that only a political system is entitled to be called democratic, that democracy is merely a mechanism for choosing and authorizing governments or in some other way getting laws and political decisions made. But we should bear in mind that democracy more often has been, and is, thought of as much more than that. From Mill through L. T. Hobhouse, A. D. Lindsay, Woodrow Wilson, and John Dewey, to the current proponents of participatory democracy, it has been seen as a quality pervading the whole life and operation of a national or smaller community, or if you like as a kind of *society*, a whole set of reciprocal relations between the people who make up the nation or other unit. Some theorists, mostly twentieth-century ones, insist on keeping the two senses separate. Some would even exclude the second sense altogether, by defining democracy as simply a system of government. But in any realistic analysis the two senses merge into each other. For different models of democracy, in the narrow sense, are congruent with, and require, different kinds of society.

Enough now has been said about models in general to indicate why an analysis of liberal democracy may conveniently be cast in terms of models. To examine models of liberal democracy is to examine what the people who want it, or want more of it, or want some variant of the present form of it, believe it is, and also what they believe it might be or should be. This is more than one can do by simply analysing the operations and institutions of any existing liberal democratic states. And this extra knowledge is important. For people's beliefs about a political system are not something outside it, they are *part* of it. Those beliefs, however they are formed or determined, do determine the limits and possible development of the system: they determine what people will put up with, and what they will demand. In short, to work in terms of models makes it easier to keep in mind that liberal democracy (like any other political system) has two necessary ingredients that may not appear on the surface: (a) to be workable, it must be not far out of line with the wants and capabilities of the human beings who are to work it; hence, the model of democracy must contain (or take for granted) a model of man; and (b), since it needs general assent and support in order to be workable, the model must contain, explicitly or implicitly, an ethically justificatory theory.

(ii) *Why historically successive models?*

If our object is to examine the limits and possibilities of contemporary liberal democracy, why should we indulge in a 'Life and Times'? Why not confine ourselves to a current analysis? Would it not be simpler to set up a single model of present liberal democracy, by listing the observable characteristics of the practice and theory common to those twentieth-century states which everyone would agree to call liberal democracies, that is, the systems in operation in most of the English-speaking world and most of Western Europe? Such a model could easily be set up. The main stipulations are fairly obvious. Governments and legislatures are chosen directly or indirectly by periodic elections with universal equal franchise, the voters' choice being normally a choice between political parties. There is a sufficient degree of civil liberties (freedom of speech, publication, and association, and freedom from arbitrary arrest and imprisonment) to make the right to choose effective. There is formal equality before the law. There is some protection for minorities. And there is general acceptance of a principle of maximum individual freedom consistent with equal freedom for others.

Many contemporary political writers do set up such a model. It can serve as a framework for investigating and displaying the actual, the necessary, and the possible workings of contemporary liberal democracy. It can also be used to argue the ethical superiority of liberal democracy over other systems. Why then should we not use a single model constructed from present practice and present theory? Why look at successive models that have prevailed in turn in the century or so down to our time?

The simplest reason is that using successive models reduces the risk of myopia in looking ahead. It is all too easy, in using a single model, to block off future paths; all too easy to fall into thinking that liberal democracy, now that we have attained it, by whatever stages, is fixed in its present mould. Indeed, the use of a single contemporary model almost commits one to this position. For a single model of current liberal democracy, if it is to be realistic as an explanatory model, must stipulate certain present mechanisms, such as the competitive party system and wholly indirect (i.e. representative) government. But to do this is to foreclose options that may be made possible by changed social and economic relations. There may be strong differences of opinion about whether some conceivable future forms of democracy can properly be called *liberal* democracy, but this is something that needs to be argued, not put out of court by definition. One of the things that needs to be considered is whether liberal democracy in a large nation-state is capable of moving to a mixture of indirect and direct democracy: that is, is capable of moving in the direction of a fuller participation, which may require mechanisms other than the standard party system.

There is another reason for preferring successive models: their use is more likely to reveal the full content of the contemporary model, the full nature of the present system. For the presently prevalent model is itself an amalgam, produced by partial rejection and partial absorption of previous models. Each of the first three models I have chosen has been for a time the prevalent model, that is, has been the one generally accepted, by those who were at all favourable to democracy, as a statement of what democracy is, what it is for, and what institutions it needs. And each successive model, after the first, was formulated as an attack on one or more of the previous models. Each has been offered as a corrective to or replacement of its predecessor: the point of departure has always been an attack on at least some part of a preceding model, even when, as has often been the case, the new model embodied substantial elements of an earlier one, sometimes without the formulators apparently being aware of this. Thus each of the models is to some extent an overlay on previous ones. So we are more likely to see the full nature of contemporary liberal democracy, and its possible future direction and limits, by looking at the successive models, and at the reasons for their creation and for their failure.

(iii) *Why these models?*

Even if we are persuaded of the merits of model-building, and of the value of analysing liberal democracy by examining successive prevalent models, the question may be asked, why choose, as I have chosen, to go back no farther than the nineteenth century? Why not go back at least to Rousseau or Jefferson, or to the democratic ideas associated with seventeenth-century Puritanism, as is more usually done by those who want to trace the roots of modern liberal democracy?

This question cannot, without circular reasoning, be settled simply by definition. One could easily put forward a definition of liberal democracy by which some pre-nineteenth-century theories and visions of democracy would qualify for inclusion. Thus if, as seems not unreasonable, one reduced the essentials of liberal democracy to three or four stipulations—say, an ideal of equal individual rights to self-development, equality before the law, basic civil liberties, and popular sovereignty with an equal political voice for all citizens—leaving out any stipulations about representation, party systems and so on, then some earlier ideas of democracy could be included as liberal democratic. Equally reasonably, by putting in stipulations about representation etc. one may exclude various earlier concepts. The definition of the model depends on value judgements about what *are* the essentials, and those judgements cannot be defended merely by invoking a definition.

Are we left, then, with no basis for choosing between possible starting-points for liberal democracy? I think not. For if our concern is with the possible future of liberal democracy, we must pay attention to the relation between democratic institutions and the underlying structure of society. And there is one such relation, largely neglected by current theorists of liberal democracy, which may be thought to be decisive. This is the relation between democracy and class.

I want now to argue that the most serious, and least examined, problems of the present and future of liberal democracy arise from the fact that liberal democracy has typically been designed to fit a scheme of democratic government onto a *class-divided* society; that this fit was not attempted, either in theory or in practice, until the nineteenth century; and that, therefore, earlier models and visions of democracy should not be counted as models of liberal democracy.

PRECURSORS OF LIBERAL DEMOCRACY

(i) *Democracy and class*

As soon as attention is focused on the relation between democracy and class, the historical record falls into a new pattern. It is, of course, not new to notice that in the main Western tradition of political thought, from Plato and Aristotle down to the eighteenth and nineteenth centuries, democracy, when it was thought of at all, was defined as rule by the poor, the ignorant, and incompetent, at the expense of the leisured, civilized, propertied classes. Democracy, as seen from the upper layers of class-divided societies, meant class rule, rule by the wrong class. It was a class threat, as incompatible with a liberal as with a hierarchical society. The main Western tradition down to the eighteenth and nineteenth centuries, that is to say, was undemocratic or anti-democratic.

But there were, indeed, in that whole stretch of over 2000 years, recurrent democratic visions, democratic advocates, and even some examples of democracy in practice (though these never embraced a whole political community). When we look at these democratic visions and theories we shall find that they have one thing in common, which sets them sharply apart from the liberal democracy of the nineteenth and twentieth centuries. This is, that they all depended on, or were made to fit, a non-class-divided society. It is hardly too much to say that for most of them democracy *was* a classless or a one-class society, not merely a political mechanism to fit such a society. These earlier models and visions of democracy were reactions against the class-divided societies of their times. As such they may properly be called utopian, an honourable name derived from the title of Thomas More's astonishing sixteenth-century work *Utopia*.

This puts them in striking contrast to the liberal-democratic tradition from the nineteenth century on, which accepted and acknowledged from the beginning—and more clearly at the beginning than later—the class-divided society, and set out to fit a democratic structure onto it.

The concept of a liberal democracy became possible only when theorists—first a few and then most liberal theorists—found reasons for believing that 'one man, one vote' would not be dangerous to property, or to the continuance of class-divided societies. The first systematic thinkers to find so were Bentham and James Mill, in the early nineteenth century. As we shall see (in Chapter II) they based that conclusion on a mixture of two things: first, deduction from their model of man (which assimilated all men to a model of bourgeois maximizing man, from which it followed that all had an interest in maintaining the sanctity of property), and second, their observation of the habitual deference of the lower to the higher classes.

So I find the watershed between utopian democracy and liberal democracy to come in the early nineteenth century. That is my reason for treating the pre-nineteenth-century theories as precursors of liberal democracy, rather than treating any of them, say Rousseau or Jefferson or any of the seventeenth-century Puritan theorists, as part of the 'classical' liberal democratic tradition. This is not to say that the pre-nineteenth-century concepts have been neglected or dismissed by the twentieth-century theorists. On the contrary, the earlier concepts have not infrequently been drawn in and appealed to, particularly by twentieth-century exponents of what I am calling Model 2. But this has not been much help to such exponents, for they have generally failed to notice that the class assumptions of the earlier theories were incongruous with their own.

I have said that those who presented favourable models or visions of democracy before the nineteenth century intended them to fit, or to be, either classless or predominantly one-class societies. Before looking at the pre-nineteenth-century record it will be well to state more specifically what is meant by class in this context.

Class is understood here in terms of property: a class is taken to consist of those who stand in the same relations of ownership or non-ownership of productive land and/or capital. A somewhat looser concept of class, defined at its simplest in terms of rich and poor, or rich and middle and poor, has been prominent in political theory as far back as one likes to go, though in the earliest theories (such as Aristotle's) the criterion of class was only implicitly ownership of *productive* property. However, the view that class, defined at least implicitly in terms of productive property, was an important criterion of different forms of government, and even an important determinant of what forms of government could come into existence and could work, was a view held by Aristotle, by Machiavelli, by the seventeenth-century English republicans, and by the American Federalists, long before Marx found in class conflict the motor of history.

Some of the non-democratic theorists who gave class a central place in their analyses (for instance, Harrington) were much concerned with distinctions between classes based not just on property or no property, but on different kinds of property relations, such as feudal versus non-feudal. But the democratic theorists generally kept their eyes on a simpler distinction: that between societies with two classes, societies with only one class, and societies with no classes. Thus, some of the earlier utopians (like the present-day communists) have envisaged a society with *no* individual ownership of productive land or capital, hence no property classes: this we may call a *classless* society. Different from this is the idea of society where there is individual ownership of productive land and capital and where everyone owns, or is in a position to own, such property: this we may call a *one-class* society. Finally there is the society where there is individual ownership of productive land and capital and where not everyone, but only one set of people, owns such property: this is the *class-divided* society.

The distinction here made between 'classless' and 'one-class' may seem somewhat arbitrary: the societies, or visions of society, I am so describing might both of them be properly enough described by either term. But since the two societies are significantly different, two different terms are needed to describe them, and it is more in accord with modern usage to keep the term 'classless' for a society with no private ownership of productive land or capital, and 'one-class' for a society where everyone does or may own such productive resources.

(ii) *Pre-nineteenth-century theories as precursors*

Let us now look at the record of democratic theory before the nineteenth century. In the ancient world there were of course some outstanding actual functioning democracies, most notably the Athens celebrated by Pericles. But no record of any substantial theory justifying or even analysing democracy has survived from that era.[1] We may surmise that any such

[1] Aristotle did briefly analyse various kinds of 'democracy', under which head he included systems with a moderate property qualification for voting. He was strongly opposed to full democracy: the only kind in which he found any merit was one in which 'husbandmen and those of moderate fortune' had supreme power (*Politics*, IV c. 6, 1292 b; cf. VI c. 4, 1318 b).

theory would have taken, as the required base for democracy, a citizen body made up mainly of persons not dependent on employment by others: that, at least, would correspond pretty well to the facts, as far as we know them, about the Athenian city-state in its democratic period, which has been well described as a property-owning democracy. We do not know if such a requirement, which amounts to the requirement of a *one-class* citizen body, was built into a theoretical model, since no theoretical model has come down to us: there can be no more than a reasonable supposition that it was.

In the Middle Ages one would not expect, nor does one find, any theory of democracy, or any demand for a democratic franchise: such popular uprisings as flared up from time to time were not concerned about an electoral franchise, for at that time power did not generally lie in elected bodies. Where feudalism prevailed, power depended on rank, whether inherited or acquired by force of arms. No popular movement, however enraged, would think that its aims could be achieved by its getting the vote. And in the nations and independent city-states of the later Middle Ages also, power was not to be sought in that way. Where voices were raised and rebellions mounted against the late medieval social order, as in the Jacquerie in Paris (1358), the uprising of the Ciompi in Florence (1378), and the Peasants' Revolt in England (1381), the demands were for levelling of ranks, and sometimes for levelling of property, rather than for a democratic political structure. They wanted either a classless communistic society, as indicated in the sentiment attributed to John Ball, of Peasants' Revolt fame: 'Things cannot go well in England, nor ever will, until all goods are held in common, and until there will be neither serfs nor gentlemen, and we shall all be equal',[2] or a levelled society where all might have property. There is no record of any of these movements having produced any systematic theory, nor having sketched a democratic political structure.

When we move on to the sixteenth and seventeenth centuries we do find some explicit democratic theories. Two democratic currents appear then in England. One of them has a classless

[2] Quoted in M. Beer: *A History of British Socialism*, London, 1929, i. 28.

base, the other a one-class base. The democratic utopias of those centuries, the best-known of which are More's *Utopia* (1516) and Winstanley's *The Law of Freedom* (1652), were classless societies. They were envisioned as replacing class-divided societies: their authors constructed them to denounce all class systems of power. Finding the basis of class oppression and exploitation in the institution of private property, they replaced it by communal property and communal work. These early modern visions of democracy were visions of a fundamentally equal, unoppressive society, as well as prescriptions for a scheme of government. Such a society had to be classless, and to be classless it had to be without private property.

The other seventeenth-century democratic current, in so far as it flowed in political and not simply religious channels, is no less related to class. English Puritanism, in that century, was rife with democratic ideas. Although these were generated by controversies about church government, and were actually put into effect only in that sphere (and, very briefly, in the army), they did spill over into ideas about civil government, especially in the period of the Civil Wars and the Commonwealth. But, except for such extreme radical utopians as Winstanley, the groups and movements whose political thinking may be said to have emerged from democratic Puritanism were not politically democratic. They did not go so far as to demand full popular sovereignty or a fully democratic franchise.

The Presbyterians and the Independents insisted on a property qualification for the franchise. About the position of the other main political movement, the Levellers, who were for a few years during the Civil Wars very strong, there is some dispute. I have shown elsewhere[3] that the Levellers, as an organized movement, speaking in concerted manifestos, intended to exclude all wage-earners and alms-takers (more than half the adult males) from the franchise. But some historians[4]

[3] *The Political Theory of Possessive Individualism*, Oxford, 1962, ch. 3; and *Democratic Theory, Essays in Retrieval*, Oxford, 1973, Essay 12.

[4] Keith Thomas: 'The Levellers and the Franchise', in G. E. Aylmer (ed.): *The Interregnum: the Quest for Settlement, 1640–1660*, London, 1972; and M. A. Barg, as cited in Christopher Hill: *The World Turned Upside Down*, London, 1972, pp. 94, 97.

have argued, in reply, that the Levellers, in their individual writings and speeches, were not unanimous about this, and that some of them were full democrats. If this is allowed as a possible interpretation of the statements of some of the Levellers, we have to ask what class structure was thought, by any democratic Levellers, to be consistent with or required by the democracy they wanted? The answer is clear. All the Levellers were strongly against the class differences they saw around them, which enabled a class of landlords and moneyed men to dominate and exploit the men of small property (and even to reduce the latter to men of no property). Some of the most vehement Leveller tracts[5] saw a class conspiracy of the men of wealth and rank, and wanted to put it down. The ideal of all the Levellers was a society where all men had enough property to work on as independent producers, and where none had the kind or amount of property which would enable them to be an exploitive class.

In short, the Levellers, whether or not any of them embraced full democracy, all cherished the ideal of a one-class society. The Levellers had the same historical view of society as Rousseau was to have a century later. They found that the rot had set in with exploitive private property. The small private property of the independent producer was a natural right. The large private property which enabled its owner to exploit the rest was a contradiction of natural right.

When we reach the eighteenth century we find some substantial theories—not many—which are usually, and quite properly, called democratic. We may take, as the leading eighteenth-century exponents of democracy, Rousseau and Jefferson: their democratic ideas have been more influential, more carried over into our own time, than any others of that century.[6] Much as Rousseau's and Jefferson's positions differed

[5] e.g. those cited in *The Political Theory of Possessive Individualism*, pp. 154–6.

[6] James Madison has no doubt been at least as influential as Jefferson, if not more so, in American thinking: Robert Dahl for instance builds his twentieth-century model of democracy largely on Madison. And Madison appears to be an exception to my generalization, for he did, in the 1780s, recognize a class-divided society, and did try to fit a system of government to it. But he is no exception, for the system he proposed can scarcely be

in other respects, both of them required a society where everyone had, or could have, enough property to work on or work with, a society of independent producers (peasants or farmers, and craftsmen), not a society divided into dependent wage-earners on the one hand, and, on the other, land and capital owners on whom they were dependent.

Rousseau's position is clear. Private property is a sacred individual right.[7] But only the moderate property of the small working proprietor is sacred. An unlimited property right, Rousseau argued forcefully in his *Discourse on the Origins of Inequality* (1755), was the source and the continuing means of exploitation and unfreedom: only a limited right was morally justifiable. He reasserted this position in *The Social Contract* (1762). The first property, property in the original means of producing the means of life, was property in a piece of land. The original right to land, the right of the first occupier, was limited in two ways: 'a man must occupy only the amount he needs for his subsistence; and . . . possession must be taken, not by any empty ceremony, but by labour and cultivation'.[8] So Rousseau found a basis in natural right for his insistence on limited property.

He needed such a limited property right for another reason, which he also made explicit: only such a limited right was consistent with the sovereignty of the general will. A truly democratic society, a society that would be governed by the

called democratic: one need only look at his anxiety to protect 'the minority of the opulent against the majority' (Max Farrand (Ed.): *The Records of the Federal Convention 1787*, revised edn., New Haven and London, 1937, i. 431); his provisions against the dominance of 'faction', which he defined as 'a number of citizens, whether a majority or a minority of the whole, who are united and actuated by some common impulse of passion or of interest' (*Federalist Papers*, No. 10); and his insistence on a natural right to unequal property, which must be protected against democratic levelling propensities (ibid.). He cannot, therefore, be enlisted as a pre-nineteenth-century liberal democrat.

[7] '. . . the right of property is the most sacred of all the rights of citizenship, and even more important in some respects than liberty itself. . . . property is the true foundation of civil society'. *Discourse on Political Economy* (1758) in *The Social Contract and Discourses* (trans. G. D. H. Cole), Everyman's Library, 1927, p. 271.

[8] Bk. I, ch. 9, in ibid., p. 20.

general will, requires such an equality of property that 'no citizen shall ever be wealthy enough to buy another, and none poor enough to be forced to sell himself'.[9] The reference to buying and selling persons is apparently not a reference to slavery, for this principle is set out as a permanent rule for *citizens*, i.e. free men: presumably, then, it is a prohibition of the purchase and sale of free wage labour. Again, 'laws are always of use to those who possess and harmful to those who have nothing: from which it follows that the social state is advantageous to men only when all have something and none too much'.[10]

Rousseau's reason for requiring such equality was clear enough. It followed directly from his insistence on the sovereignty of the general will. For where differences of property divide men into classes with opposed interests, men will be guided by class interests, which are, *vis-à-vis* the whole society, particular interests; so they will be incapable of expressing a general will for the common good. The emergence and steady operation of the general will required a one-class society of working proprietors. Such a society was to be achieved by government action: 'It is therefore one of the most important functions of government to prevent extreme inequality of fortunes; not by taking away wealth from its possessors, but by depriving all men of means to accumulate it; not by building hospitals for the poor, but by securing the citizens from becoming poor.'[11]

When we turn to the theorist who is often accounted the first great American proponent of democracy we find a similar, though less systematic, argument. Thomas Jefferson treated the common people as trustworthy to an extent unusual in most subsequent Presidents of the United States. It would be unduly cynical to think that this was because he was without the temptations afforded by modern techniques of presidential public relations. In any case, he made it clear, both in his public statements and his private letters, that his trust in the people was trust in the independent worker-proprietor,

[9] Bk. II, ch. 11, in ibid., p. 45.
[10] Bk. I, ch. 9, in ibid., p. 22, n. 1.
[11] *Discourse on Political Economy*, in ibid., p. 267.

whom he saw as the backbone, and hoped would remain the backbone, of American society.

In his most substantial published work, the *Notes on Virginia* (1791), he was clear that his favourable estimate of human nature was confined to those who had substantial economic independence:

Dependence begets subservience and venality, suffocates the germ of virtue, and prepares fit tools for the designs of ambition . . . generally speaking, the proportion which the aggregate of the other classes of citizens bears in any State to that of its husbandmen, is the proportion of its unsound to its healthy parts, and is a good enough barometer whereby to measure the degree of its corruption . . . The mobs of great cities add just so much to the support of pure government, as sores do to the strength of the human body.[12]

The same principle is expressed in a letter to John Adams in 1813:

Here everyone may have land to labor for himself, if he chooses; or, preferring the exercise of any other industry, may exact for it such compensation as not only to afford a comfortable subsistence, but wherewith to provide for a cessation from labor in old age. Everyone, by his property or by his satisfactory situation, is interested in the support of law and order. And such men may safely and advantageously reserve to themselves a wholesome control over their public affairs, and a degree of freedom, which, in the hands of the *canaille* of the cities of Europe, would be instantly perverted to the demolition and destruction of everything public and private.[13]

Democracy, for Jefferson, required a society in which everyone was independent economically. Reasoning from the American situation, Jefferson did not require that everyone should be a worker-proprietor, but only that everyone could be one if he wished. He had no objection to wage-labour, but only because, with free land available, wage-earners were as independent as husbandmen. Nor did he object to some men, like himself, having substantial estates, provided that everyone else had, or could have, a small estate sufficient to make him independent. In the circumstances which Jefferson saw prevailing in America, and which he considered prerequisite for democracy any-

[12] *Notes on Virginia*, Query XIX, in Saul K. Padover: *The Complete Jefferson*, New York, 1943, pp. 678–9.

[13] Ibid., pp. 285–6.

where, there was, therefore, no fundamental class division. He allowed the existence of a wage-relation only because it did not, in those circumstances, make a class-divided society. Jefferson's prerequisite for a democracy was, like Rousseau's, a one-class society.

It may be objected that the kind of society envisaged by these pre-nineteenth-century democratic writers as a prerequisite of democracy was not after all a one-class society, in that it would still leave women as a subordinate class, unable to own productive property in their own right. Moreover, as we have seen, the point emphasized by the democratic opponents of class-divided society was that any class without productive property was dependent on and exploited by the class with such property. It may well be argued that women were in just that position, and certainly the early democratic writers were not conspicuous for taking any stand against it: Rousseau indeed thought that women ought to be kept dependent. Were not these writers, then, assuming what must be called a class-divided society?

I think not. For down to the nineteenth century women were commonly considered not full members of society. They were in, but not of, civil society. It would scarcely occur to a theorist, in describing or prescribing the class character of a society, to treat them as a class. An eighteenth-century democrat could think of a one-class society excluding women as easily as an ancient Athenian democrat could think of a one-class society excluding slaves.

Nor can women be said to have been a class in any full sense. True, in so far as women could not own property they meet our minimum definition of a class. And in so far as they were kept dependent and exploited they fit the underlying concept of class as an exploited/exploiter relation. But there is a very great difference between the way they were exploited and the way the propertyless working class (who were also considered in the seventeenth and eighteenth centuries to be not full members of civil society[14]) were exploited. The difference is I think so great as to make it inappropriate to describe women as a class.

[14] Cf. *The Political Theory of Possessive Individualism*, pp. 221–9.

For from the seventeenth century on, as the capitalist market relation replaced feudal or other status relations as the means by which owners benefited from the work of non-owners, it was understood that the only permissible arrangement for such benefit was the relation between free wage-earners and owners of the capital which employed them. The wage relation, a strictly market relation, became the criterion of class. And in the eighteenth century, when Rousseau and Jefferson were stipulating a one-class society, and for some time after that, women were not a class by that criterion. They were indeed exploited by the male-dominated society, which made most of them perform the function of reproducing the labour force for no more reward than their subsistence. But they were made to do this by legal arrangements akin to a feudal (or even slave) relation, rather than by a market relation. In so far as class was, and was seen to be, determined by the capitalist market relation, women as such were not, and would not be thought to be, a class. That being so, writers who inveighed against class-divided society while not treating women as a class, were genuinely stipulating a one-class society. We are therefore, I think, still entitled to refer to the pre-nineteenth-century democratic theorists as advocates of a one-class (or classless) society.

This brief survey of models of democracy earlier than the nineteenth century is, I hope, sufficient to sustain my generalization that all of them were fitted either to a classless or to a one-class society. And that is why I think that all of the pre-nineteenth-century democratic theories are better treated as being outside the liberal-democratic tradition. To be counted in that tradition a theory should surely be both democratic and liberal. But what is usually, and I think rightly, considered to be the liberal tradition, stretching from Locke and the Encyclopédistes down to the present, has from the beginning included an acceptance of the market freedoms of a capitalist society.

The pattern is clear enough. The seventeenth- and eighteenth-century liberals, who were not at all democratic (from, say, Locke to Burke) fully accepted capitalist market relations. So did the early nineteenth-century liberal-democrats, how strongly in the cases of Bentham and James Mill we shall see in Chapter II. Then from about the middle of the nineteenth century to the middle of the twentieth, as we shall see in Chapter III, the liberal-democratic thinkers tried to combine an acceptance of the capitalist market society with a humanist ethical position. This produced a model of democracy notably different from Bentham's, but still including acceptance of the market society. Since the liberal component of liberal democracy has pretty constantly included acceptance of capitalist relations and hence of class-divided society, it seems appropriate that the pre-nineteenth-century democratic theories, all of which rejected the class-divided society, should be placed outside the liberal-democratic category. They were, so to speak, handicraft models of democracy, and as such are best considered as precursors of liberal democracy.

If this is thought to be still a somewhat arbitrary division, I shall not insist. The important thing is not the classification, but the recognition of how deeply the market assumptions about the nature of man and society have penetrated liberal-democratic theory.

The reader may wonder whether the grounds offered for this classification do not commit the author to the proposition that liberal democracy must always embrace the capitalist market society with its class-division. If 'liberal' has always meant that, or at least has always included that, should it continue to be used only with that meaning? Is it not then inconsistent to go on to inquire, as I do in Chapter V, into the prospects of a democratic theory which downgrades or abandons the market assumptions, and to treat this as an inquiry into a possible future model of liberal democracy?

I do not think any of these questions are to be answered in the affirmative. I would argue that the reason 'liberal' did mean acceptance of the capitalist market society, during the formative century of liberal democracy, does not apply any longer. Liberalism had always meant freeing the individual from the outdated restraints of old established institutions. By the time liberalism emerged as liberal democracy this became a claim to free all individuals equally, and to free them to use and develop their human capacities fully. But so long as there was an economy of scarcity, it still seemed to the liberal democrat that the only way to that goal was through the productivity of free-enterprise capitalism. Whether this was in fact the only way as late as the early twentieth century may be doubted, but there is no doubt that the leading liberal democrats thought it to be so; and as long as they did, they had to accept the linkage of market society with liberal-democratic ends. But this linkage is no longer necessary. It is no longer necessary, that is to say, if we assume that we have now reached a technological level of productivity which makes possible a good life for everybody without depending on capitalist incentives. That assumption may of course be challenged. But if it is denied, then there seems no possibility of any new model of democratic society, and no point in discussing such a model under any designation, liberal or otherwise. If the assumption is granted, the previously necessary linkage is no longer necessary, and a new model not based on the capitalist market may properly be considered under the heading 'liberal-democratic'.

In the following chapters I shall examine three successive models of liberal democracy that may be said to have prevailed in turn from the early nineteenth century to the present, and shall go on to consider the prospects of a fourth. The first model I call *Protective Democracy*: its case for the democratic system of government was that nothing less could in principle protect the governed from oppression by the government. The second is called *Developmental Democracy*: it brought in a new moral dimension, seeing democracy primarily as a means of individual self-development. The third, *Equilibrium Democracy*, abandoned the moral claim, on the ground that experience of the actual operation of democratic systems had shown that the developmental model was quite unrealistic: the equilibrium theorists offered instead a description (and justification) of democracy as a competition between élites which produces equilibrium without much popular participation. This is the presently prevalent model. Its inadequacy is becoming increasingly apparent, and the possibility of replacing it with something more participatory has become a lively and serious issue. So this study goes on to consider the prospects and problems of a fourth model, *Participatory Democracy*.

II

Model 1: Protective Democracy

THE BREAK IN THE DEMOCRATIC TRADITION

Whatever may be thought of Tennyson's lines about freedom slowly broadening down from precedent to precedent, it is clear that this is not the way we reached our present liberal democracies. It is true that in the present liberal democracies the universal franchise did generally come by stages, starting from a restrictive property qualification, moving at different speeds in different countries to manhood suffrage, and finally including women suffrage. But before this expansion of the franchise had begun at all, the institutions and ideology of liberal individualism were firmly established. The only apparent exceptions to this rule were no exceptions. Some European countries, notably France, did have manhood franchise before the liberal market society had fully established itself there. But since the assemblies elected by that franchise did not have the power to make or unmake governments, the arrangements cannot be deemed democratic: the extent of the franchise is a measure of democratic government only in so far as the exercise of the franchise can make and unmake governments. So we may say that by the time the movement for a fully democratic franchise had gathered momentum anywhere, the concept of democracy which that franchise was to embody was very different from any of the earlier visions of democracy.

Thus there is a sharp break in the path from pre-liberal to liberal democracy. A fresh start was made in the nineteenth century, from a very different base. The earlier concepts of democracy, as we have seen, had rejected class division, believing or hoping that it could be transcended, or even assuming that in some places—Rousseau's Geneva or Jefferson's America—it had been transcended. Liberal democracy, on the contrary, accepted class division, and built on it. The first formulators of liberal democracy came to its advocacy through a chain of reasoning which started from the assumptions of a capitalist market society and the laws of classical political economy. These gave them a model of man (as maximizer of utilities) and a model of society (as a collection of individuals with conflicting interests). From those models, and one ethical principle, they deduced the need for government, the desirable functions of government, and hence the desirable system of choosing and authorizing governments. To see how deeply their models of man and society got into their general theory, and hence into their model of liberal democracy as the best form of government, we shall do well to look more closely than is usually done at the theories of the two earliest systematic exponents of liberal democracy, Jeremy Bentham and James Mill.[1]

We may start with Bentham, the original systematizer of the theory that came to be known as Utilitarianism, and bring in James Mill when, as sometimes happened, he stated the Utilitarian case more clearly than Bentham, or when his reservations and ambiguities were different from Bentham's. James Mill was a thorough disciple of Bentham, and a much more disciplined writer, so he often put the Benthamite case more strikingly than the master himself. And by the time Bentham

[1] James Mill's model can be dated precisely at 1820, in his famous article on *Government*. Bentham's may be dated 1820 (see p. 35, n. 22) or 1818, when he produced the twenty-six *Resolutions on Parliamentary Reform*, which would admit to the franchise 'all such persons as, being of the male sex, of mature age, and of sound mind, shall . . . have been resident either as householders or inmates, within the district or place in which they are called upon to vote'. (*Works*, ed. Bowring, Edinburgh and London, 1843, x. 497.)

Others, indeed, had advocated equal manhood suffrage somewhat earlier, notably Major John Cartwright, as early as 1776, in his *Take Your Choice!*, and Cobbett in his *Political Register*. But neither of them can be said to have set up a fully reasoned model, and such theoretical grounds as they did offer were backward-looking: their appeal was to the natural rights of freeborn Englishmen (before the restrictions of the franchise by 8 Henry VI, c. 7); and there was no awareness of the changed class structure or of the significance of the new industrial working class.

put his mind to the question of the best form of government, their minds ran in parallel, and they were in close touch with each other. So it will do no injustice to either to treat them almost as a unit.

It must be said that with Bentham and James Mill liberal democracy got off to a poor start. It is not that they were incompetent theorists. On the contrary, Bentham became deservedly famous as a thinker, and the most influential doctrine of the English nineteenth century was named after him. And James Mill, though not of the very first rank, was a clear and forceful writer. And the general theory of Utilitarianism, from which they both deduced the need for a democratic franchise, seemed both fundamentally egalitarian and thoroughly businesslike. It was both, and that was the trouble. I shall suggest that it was the combination of an ethical principle of equality with a competitive market model of man and society that logically required both thinkers to conclude in favour of a democratic franchise, but made them do so either ambiguously or with reservations.

THE UTILITARIAN BASE

The general theory was clear enough. The only rationally defensible criterion of social good was the greatest happiness of the greatest number, happiness being defined as the amount of individual pleasure minus pain. In calculating the aggregate net happiness of a whole society, each individual was to count as one. What could be more egalitarian than that as a fundamental ethical principle?

But to it were added certain factual postulates. Every individual by his very nature seeks to maximize his own pleasure without limit. And although Bentham set out a long list of kinds of pleasure, including many non-material ones, he was clear that the possession of material goods was so basic to the attainment of all other satisfactions that it alone could be taken as the measure of them all. 'Each portion of wealth has a corresponding portion of happiness.'[2] And again: 'Money is

[2] *Principles of the Civil Code*, Part I, ch. 6, in Bentham: *The Theory of Legislation*, ed. C. K. Ogden, London, 1931, p. 103. (I have preferred this

the instrument of measuring the quantity of pain or pleasure. Those who are not satisfied with the accuracy of this instrument must find out some other that shall be more accurate, or bid adieu to politics and morals.'[3]

So each seeks to maximize his own wealth without limit. One way of doing this is to get power over others. 'Between wealth and power, the connexion is most close and intimate; so intimate, indeed, that the disentanglement of them, even in the imagination, is a matter of no small difficulty. They are each of them respectively an instrument of production with relation to the other.'[4] And again, 'human beings are the most powerful instruments of production, and therefore everyone becomes anxious to employ the services of his fellows in multiplying his own comforts. Hence the intense and universal thirst for power; the equally prevalent hatred of subjection.'[5]

James Mill was even more forthright. In his 1820 article *Government*, he wrote:

> That one human being will desire to render the person and property of another subservient to his pleasures, notwithstanding the pain or loss of pleasure which it may occasion to that other individual, is the foundation of government. The desire of the object implies the desire of the power necessary to accomplish the object. The desire, therefore, of that power which is necessary to render the persons and properties of human beings subservient to our pleasures is a grand governing law of human nature . . . The grand instrument for attaining what a man likes is the actions of other men. Power . . . therefore, means security for the conformity between the will of one man and the acts of other men. This, we presume, is not a proposition which will be disputed.[6]

With this grand governing law of human nature, society is a collection of individuals incessantly seeking power over and at the expense of each other. To keep such a society from flying apart, a structure of law both civil and criminal was seen to be

---

edition to the version printed in the Bentham *Works* edited by Bowring, vol. i.) On the abstraction from reality required to assert this proposition, see below, p. 30, at n. 12.

[3] W. Stark (ed.): *Jeremy Bentham's Economic Writings*, i. 117.

[4] *Constitutional Code*, Bk. 1, ch. 9, in *Works*, ed. Bowring, ix. 48.

[5] Stark (ed.): iii. 430.

[6] Section IV (p. 17 of the Barker edition, Cambridge, 1937).

needed. Various structures of law might be capable of providing the necessary order, but, of course, according to the Utilitarian ethical principle, the best set of laws, the best distribution of rights and obligations, was that which would produce the greatest happiness of the greatest number. This most general end of the laws could, Bentham said, be divided into four subordinate ends: 'to provide subsistence; to produce abundance; to favour equality; to maintain security.'[7]

BENTHAM'S ENDS OF LEGISLATION

Bentham's arguments as to how each of these ends could be achieved (and how not) are revealing. Together they amount to a case for a system of unlimited private property and capitalist enterprise, and this apparently deduced from the factual postulates about human nature and a few others. Let us look in turn at his arguments under each head.

First, subsistence. The law need do nothing to ensure that enough will be produced to provide subsistence for everyone.

> What can the law do for subsistence? Nothing directly. All it can do is to create *motives*, that is, punishments or rewards, by the force of which men may be led to provide subsistence for themselves. But nature herself has created these motives, and has given them a sufficient energy. Before the idea of laws existed, *needs* and *enjoyments* had done in that respect all that the best concerted laws could do. Need, armed with pains of all kinds, even death itself, commanded labour, excited courage, inspired foresight, developed all the faculties of man. Enjoyment, the inseparable companion of every need satisfied, formed an inexhaustible fund of rewards for those who surmounted obstacles and fulfilled the end of nature. The force of the physical sanction being sufficient, the employment of the political sanction would be superfluous.[8]

What the laws can do is to 'provide for subsistence indirectly, by protecting men while they labour, and by making them sure of the fruits of their labour. Security for the labourer, security for the fruits of labour; such is the benefit of laws; and it is an inestimable benefit.'[9]

[7] *Principles of the Civil Code*, Part I, ch. 2; Ogden (ed.): op. cit., p. 96.
[8] Ibid., Part I, ch. 4; Ogden, p. 100.
[9] Ibid.

The curious point here is that Bentham, in invoking fear of starvation as a natural incentive to the productive labour which would provide subsistence for everybody, has slipped from thinking of a primitive society ('before the idea of laws existed'), where fear of starvation would have that effect on everybody, to an advanced nineteenth-century industrial society, where that does not apply without an additional proviso. In a primitive society with such a low level of productive technique that the incessant labour of all was needed (and was seen by all to be needed) to avoid general starvation, the fear of starvation would be a sufficient incentive to the productive labour that would produce subsistence for all. But in a society whose productive techniques are sufficient to provide subsistence for everyone without such incessant labour by everyone, like England in Bentham's time, fear of starvation is not in itself a sufficient incentive. In such a society, fear of starvation will be an incentive to incessant labour only where the institutions of property have created a class who have no property in land or working capital, and no claims on society for their support, and hence must sell their labour or starve.

So keen a thinker as Bentham could scarcely have failed to see this, had he not been taking for granted the existence of such a class as inevitable in any economically advanced society. And we know the he did assume this: 'In the highest state of social prosperity, the great mass of citizens will have no resource except their daily industry; and consequently will be always near indigence.'[10] Already we can see the teachings of classical political economy subverting the egalitarian principle.

A similar shift takes place in his argument about 'abundance'. Here he seems to slip from thinking of a society of independent producers to thinking of his own advanced society, applying to the latter a generalization about incentives apparently drawn from the former. No legislation, he says, is needed to encourage individuals to produce abundance of material goods. Natural incentives are enough, because everyone's desire is infinite. Each want satisfied produces a new want. So there is a strong and permanent incentive to produce more.

[10] Ibid., Part I, ch. 14; Ogden, p. 127.

Bentham does not notice that this incentive, which may properly enough be postulated of the capitalist entrepreneur and possibly of the self-employed independent producer, cannot very well apply to the wage-earners, who are 'always near indigence'. He does not see this, because he has created his model of man in the image of the entrepreneur or the independent producer. He could do that because he had no historical sense.

It is only when we come to his argument under the heads of equality and security that we can see the full extent to which his acceptance of capitalism undermined his egalitarian ethical principle. The case for 'equality', that is, for everyone having the same amount of wealth or income, is set out clearly. It rests on what came to be known as the law of diminishing utility, which points out that successive increments of wealth (or of any material goods) bring successively less satisfaction to their holder, or, that a person with ten or a hundred times the wealth of another has much less than ten or a hundred times as much pleasure. Given that all individuals have the same capacity for pleasure, and that 'each portion of wealth has a corresponding portion of happiness', it follows that 'he who has the most wealth has the most happiness', but also that 'the excess in happiness of the richer will not be so great as the excess of his wealth'.[11] From this it follows that aggregate happiness will be greater the more nearly the distribution of wealth approaches equality: maximum aggregate happiness requires that all individuals have equal wealth.

This case for equality requires, as we have noticed, an assumption of equal capacities for pleasure. For if some were assumed to have a greater capacity for pleasure, i.e. a greater sensitivity or sensibility, it could be argued that aggregate happiness would be maximized by their having more wealth than the others. Bentham was not very consistent about this. He prefaced the 'diminishing returns' argument for equality by setting aside 'the particular sensibility of individuals, and . . . the exterior circumstances in which they may be placed'. These must be set aside, he said, because 'they are never the same for two individuals', so that, without setting those differences aside, 'it will be impossible to announce any general

[11] Ibid., Part I, ch. 6; Ogden, p. 103.

proposition'.[12] Yet elsewhere he pointed out that, besides particular individual differences in sensibility, there were differences between whole categories of individuals. There was a difference in sensibility as between the sexes: 'In point of quantity, the sensibility of the female sex appears in general to be greater than that of the male.'[13] And, of more direct importance in an argument that depends on a relation between pleasure and wealth, Bentham saw a difference in sensibility between those of different 'station, or rank in life': '*Caeteris paribus*, the quantum of sensibility appears to be greater in the higher ranks of men than in the lower.'[14] If Bentham had acknowledged such a property-class differential when making his case for equality of wealth, his case would have been destroyed: he would have been endorsing the position of Edmund Burke. Perhaps he was. Perhaps he saw no need to mention that differential when stating his case for equality because he had already decided that the claims of equality were entirely subordinate to the claims of security.

In any case, having said this much under the head of 'equality', Bentham turned to 'security', that is, security of property and of expectation of return from the use of one's labour and property. Without security of property in the fruits of one's labour, Bentham says, civilization is impossible. No one would form any plan of life or undertake any labour the product of which he could not immediately take and use. Not even simple cultivation of the land would be undertaken if one could not be sure that the harvest would be one's own. The laws, therefore, must secure individual property. And since men differ in ability and energy, some will get more property than others. Any attempt by the law to reduce them to equality would destroy the incentive to productivity. Hence, as between equality and security, the law must have no hesitation: 'Equality must yield.'[15]

The argument is persuasive, though invalid. True, if one

[12] Ibid.

[13] *Introduction to the Principles of Morals and Legislation*, ch. 6, in *Collected Works*, London, 1970, p. 64.

[14] Ibid., p. 65.

[15] *Principles of the Civil Code*, Part I, ch. 11; Ogden, p. 120.

accepts Bentham's premiss that every individual by his very nature seeks to maximize his pleasure, and hence his material goods, without limit, and at the expense of others, it *does* follow that security for the fruits of one's labour is needed to convert the search for gain into an incentive to produce. But it does *not* follow, as Bentham argued, that no society above savagery is possible without that security, unless security for the fruits of one's labour is stretched to include the security of subsistence enjoyed by the slaves in ancient high civilizations. Forced labour, whether in the form of slavery or in any other form, is quite capable of sustaining a high level of civilization; and on Bentham's own premiss that everyone seeks power over others because 'human beings are the most powerful instruments of production', he could scarcely rule this out as unnatural. In fact, as we shall see in a moment, rather than ruling it out he endorses it.

However, if he had been content to limit his case for security of property to the case for security for the fruits of one's labour, he would have had a fairly effective case. But he was not content with that. He made another of his unconscious shifts. He went on to a very different proposition: that security of any existing kind of established property, including that which could not possibly be the fruits of one's own labour, must be guaranteed.

In consulting the grand principle of security what ought the legislator to decree respecting the mass of property already existing?

He ought to maintain the distribution as it is actually established. . . . There is nothing more different than the state of property in America, in England, in Hungary, and in Russia. Generally, in the first of these countries, the cultivator is a proprietor; in the second, a tenant; in the third, attached to the glebe; in the fourth, a slave. However, the supreme principle of security commands the preservation of all these distributions, though their nature is so different, and though they do not produce the same sum of happiness.[16]

Bentham's supporting argument demonstrates again his lack of historical sense. His contention is, that to overturn *any* existing system of property is to make impossible any other system

[16] Ibid., Part I, ch. 11; Ogden, p. 119.

of property. It does not need a profound knowledge of history to see that this is not so. For instance, the destruction of the feudal system of property led to the establishment of an equally firm capitalist system of property; and the same might be said of many previous overthrows of an existing system.

If Bentham's unhistorical postulate had been true, he would have been logically entitled to conclude that every established system must be maintained, even where it did not 'produce the same sum of happiness'; for the overturning of any system would then be worse, by the greatest happiness criterion, than any possible benefit from another system. But the postulate is not valid. So his 'demonstration' that security has absolute priority over equality is not valid.

It might be thought that Bentham could have established his case for the security of *any* established system of property, including those which maintained an extremely unequal distribution of wealth, without relying on his unhistorical postulate but simply by invoking another principle which he announced in the chapter on equality. This is the principle that

> men in general appear to be more sensitive to pain than to pleasure, even when the cause is equal. To such a degree, indeed, does this extend, that a loss which diminishes a man's fortune by one-fourth, will take away more happiness than he could gain by doubling his property.[17]

But Bentham saw that this alone did not justify the maintenance of great inequality. All he concluded from this was that, as between two persons of *equal* wealth, a redistribution would mean a net loss of happiness. He could have shown further, that as between two persons one of whom started with four times the wealth of another, a redistribution of a quarter of A's wealth to B, which would double B's wealth, would still mean some net loss of happiness. But if A started with say, twelve, times the wealth of B, a redistribution of a quarter of A's wealth would quadruple B's wealth, which presumably would mean a net gain in happiness. Bentham recognized this. His way of putting it was to say that in such a case 'the evil done by an attack on security will be compensated in part by a good which

[17] Ibid., Part I, ch. 6; Ogden, p. 108.

will be great in proportion to the progress towards equality'.[18] So he needed an independent argument to make his case for the absolute priority of security over equality. And the independent argument was, as we have seen, based on the invalid historical postulate.

From Bentham's whole treatment of the four subordinate ends of legislation, and from his preceding factual postulates, it is clear, then, how deeply his general theory was penetrated by bourgeois assumptions. First we have the general postulates: that every person always acts to secure his own interest, to maximize his own pleasure or utility, without limit; and that this conflicts with everyone else's interest. Then the search for the maximum pleasure is reduced to the search for maximum material goods and/or power over others. Then, postulates drawn from his contemporary capitalist society are presented as universally valid: that the great mass of men will never rise above a bare subsistence level; that for them fear of starvation rather than hope of gain is the operative incentive to labour; that, for the more fortunate, hope of gain is a sufficient incentive to maximum productivity; that, for this hope to operate as an incentive, there must be absolute security of property. Finally, we have security of property elevated to a 'supreme principle' absolutely overriding the principle of equality.

The ultimate reason Bentham saw no contradiction here, the reason underlying his unhistorical postulate, is, I suggest, that he was really concerned only with the rationale of the capitalist market society. In that society indeed, at least according to his version of classical political economy, there appeared to be no such contradiction: security of unlimited individual appropriation was the very thing which, along with unlimited desire, would induce the maximum productivity of the whole system. But to say that security of property, while perpetuating inequality, maximizes productivity, is not to say that it maximizes aggregate pleasure or utility. Bentham has again shifted his ground, now from aggregate utility to aggregate wealth. But these are different. The shift is illegitimate because, by his own principle of diminishing utility, a smaller national wealth, equally distributed, could yield a larger aggregate utility than

[18] Ibid.

a larger national wealth unequally distributed. But Bentham was so imbued with the ethos of capitalism, which is for maximization of wealth and sees it as equivalent to maximization of utility, that he did not admit their difference.

THE POLITICAL REQUIREMENT

For this kind of society, what kind of state was needed? The political problem was to find a system of choosing and authorizing governments, that is, sets of law-makers and law-enforcers, who would make and enforce the kind of laws needed by such a society. It was a double problem: the political system should both produce governments which would establish and nurture a free market society and protect the citizens from rapacious governments (for by the grand governing principle of human nature every government would be rapacious unless it were made in its own interest not to be so, or impossible for it to be so).

The crucial point in the solution of this double problem turned out to be the extent of the franchise, along with certain devices such as the secret ballot, frequent elections, and freedom of the press, which would make the vote a free and effective expression of the voter's wishes. The extent and genuineness of the franchise became the central question because, by the early nineteenth century in England, theorists were able to take for granted the rest of the framework of representative government: the constitutional provisions whereby legislatures and executives were periodically chosen, and therefore periodically replaceable, by the voters at general elections, and whereby the civil service (and the military) were subordinate to a government thus responsible to the electorate. So the model which the nineteenth-century thinkers started from was a system of representative and responsible government of this kind. The question that was left for them was, what provisions for the extent and genuineness of the franchise would both produce governments which would promote a free market society and protect the citizens from the government.

If only the first of these requirements had been seen as a problem, something far short of a democratic franchise would have been sufficient. Indeed, something far short of that satisfied Bentham for two decades after he began to think about political systems. In a work written between 1791 and 1802 he was for a limited franchise, excluding the poor, the uneducated, the dependent, and women.[19] In 1809 he was advocating a householder franchise, one limited to those paying direct taxes on property.[20] By 1817 he was talking about a 'virtually universal' franchise, excluding only those under age and those unable to read, and possibly excluding women (to give a decided opinion on that 'would be altogether premature in this place'); but in that same work he said that while he had become convinced of the safeness of the principle of universal suffrage, he was also convinced 'of the ease and consistency with which, for the sake of *union* and *concord*, many exclusions might be made, at any rate for a time and for the sake of quiet and gradual experience.'[21] By 1820 he was for manhood franchise; but even then he said that he would gladly support the more limited householder franchise except that he could not see that this could satisfy those excluded, who 'would perhaps constitute a majority of male adults'.[22] So Bentham was not enthusiastic about a democratic franchise: he was pushed to it, partly by his appraisal of what the people by then would demand, and partly by the sheer requirements of logic as soon as he turned his mind to the constitutional question.

'Every body of men [including whatever body has the power to legislate and to govern] is governed altogether by its conception of what is its interest, in the narrowest and most selfish sense of the word interest: never by any regard for the interest of others.'[23] The only way to prevent the government despoiling all the rest of the people is to make the governors frequently removable by the majority of all the people. The powers of government in the hands of any set of people other than those chosen and removable by the votes of the greatest number

[19] *Principles of Legislation*, ch. 13, sect. 9; in Ogden (ed.): *The Theory of Legislation*, p. 81.

[20] *Plan of Parliamentary Reform*, 1818 edn., pp. 40 n. and 127.

[21] Ibid., pp. 35–7 and 40 n.

[22] *Radicalism Not Dangerous*, in *Works*, ed. Bowring, iii. 599.

[23] *Constitutional Code*, in *Works*, ed. Bowring, ix. 102.

'would be necessarily directed to the giving every possible increase to their own happiness, whatever became of the happiness of others. And in proportion as their happiness received increase would the aggregate happiness of all the governed be diminished.'[24] Happiness is a zero-sum game: the more the governors have, the less the governed have.

The case for a democratic system is purely the protective case: 'with the single exception of an aptly organized democracy, the ruling and influential few are enemies of the subject many: . . . and by the very nature of man . . . perpetual and unchangeable enemies.'[25]

A democracy, then, has for its characteristic object and effect, the securing its members against oppression and depredation at the hands of those functionaries which it employs for its defence . . .

Every other species of government has necessarily, for its characteristic and primary object and effect, the keeping the people or non-functionaries in a perfectly defenceless state, against the functionaries their rulers; who being, in respect of their power and the use they are disposed and enabled to make of it, the natural adversaries of the people, have for their object the giving facility, certainty, unbounded extent and impunity, to the depredation and oppression exercised on the governed by their governors.[26]

But while logical deduction from the nature of human beings gave an irrefutable case for a democratic constitution, Bentham was ready to compromise it on grounds of expediency. His final position on female suffrage is a clear example. The case for universal franchise required that women, equally with men, should have the vote. Indeed, Bentham argued that, to compensate for their natural handicaps, women were if anything entitled to more votes than men. Nevertheless, he held that there is now such a general presupposition against female suffrage that he could not recommend it: 'the contest and confusion produced by the proposal of this improvement would entirely engross the public mind, and throw improvement, in all other shapes, to a distance.'[27]

[24] Ibid., p. 95.
[25] Ibid., p. 143.
[26] Ibid., p. 47.
[27] Ibid., p. 109.

So we have Bentham's whole position on the democratic franchise. He would be happy with a limited franchise but was willing to concede manhood franchise. In principle he even made a case for universal franchise, but held that the time was not ripe for it: to advocate votes for women now would endanger the chances of any parliamentary reform. And we should notice that he moved to the principle of the democratic franchise only when he had become persuaded that the poor would not use their votes to level or destroy property. The poor, he argued, have more to gain by maintaining the institution of property than by destroying it, and as evidence he pointed to the fact that in the United States those 'without property sufficient for their maintenance' had, for upwards of fifty years, 'had the property of the wealthy within the compass of their legal power' and had never infringed property.[28]

JAMES MILL'S SEESAW

It was James Mill who, in 1820, made the most powerful case for universal franchise, and even that was so guarded and put in such hypothetical terms that it can be read, and often has been read, as a case for a much less than universal franchise.[29] But though he hedged his conclusions, his argument leads irresistibly to universal franchise. The main argument is bolder than Bentham's but essentially similar. It starts with the assertion of what is surely the most extreme postulate about self-interest ever made, before or since—that grand governing law of human nature that we have already seen. From this it followed that those who had no political power would be oppressed by those who did have it. The vote was political power, or at least the lack of the vote was lack of political power. Therefore everyone needed the vote, for self-protection. Nothing short of 'one person, one vote' could in principle protect all the citizens from the government.

[28] Ibid., p. 143.

[29] The various readings are discussed by Joseph Hamburger: 'James Mill on Universal Suffrage and the Middle Class', *Journal of Politics* (1962), vol. 24, pp. 167–90; and in Hamburger: *Intellectuals in Politics, John Stuart Mill and the Philosophic Radicals*, New Haven and London, 1965, pp. 48–53.

But it cannot be said that James Mill was enthusiastic about democracy, any more than was Bentham. For in the same article on *Government* in which he made the case for a universal franchise, James Mill used considerable ingenuity in enquiring whether any narrower franchise could give the same security to every citizen's interest as would universal franchise, and he argued that it would be safe to exclude all women, all men under the age of 40, and the poorest one-third of the males over 40.

The argument is almost unbelievably crude. His general principle was that 'all those individuals whose interests are indisputably included in those of other individuals may be struck off without inconvenience'.[30] That seems fair enough, but his applications of the principle were brusque and overbearing. In the first place, Mill held, this took care of women, 'the interest of almost all of whom is involved either in that of their fathers or in that of their husbands'.[31] It also permitted the exclusion of all males under some assigned age, about which age 'considerable latitude may be taken without inconvenience. Suppose the age of forty were prescribed . . . scarcely any laws could be made for the benefit of all the men of forty which would not be laws for the benefit of all the rest of the community.' And 'the great majority of old men have sons, whose interest they regard as part of their own. This is a law of human nature. There is, therefore, no great danger that, in such an arrangement as this, the interests of the young would be greatly sacrificed to those of the old.'[32] (Mill was 47 in 1820.)

When it came to the question of an allowable property or income qualification, Mill did not even try to apply his principle of included interests. The question Mill posed was whether, somewhere between a qualification so low as to be of no use and one so high as to constitute an undesirable aristocracy of wealth, there is one 'which would remove the right of Suffrage from the people of small, or of no property, and yet constitute an elective body, the interest of which would be identical with

[30] *An Essay on Government,* ed. E. Barker, Cambridge, 1937, p. 45.

[31] Ibid., p. 45.

[32] Ibid., pp. 46–7.

that of the community?'[33] Although this is posed as a question of identity of interests, the answer is in terms of a calculation of opposed interests. Mill's answer is that a property qualification high enough to exclude up to one-third of the people (presumably one-third of the males over 40) would be safe, because each of the top two-thirds, who would have the vote, and who would of course have an interest in oppressing the excluded one-third, 'would have only one-half the benefit of oppressing a single man. In that case, the benefits of good Government, accruing to all, might be expected to overbalance to the several members of such an elective body the benefits of misrule peculiar to themselves. Good Government would, therefore, have a tolerable security.'[34] By the same token, a property qualification which excluded more than half of the people was undesirable, for it would mean that each voter 'would have a benefit equal to that derived from the oppression of more than one man':[35] this benefit would be irresistible, so that bad government would be ensured.

We can scarcely avoid asking why James Mill, after making his strong positive case for universal suffrage, should have raised the question of exclusions at all, let alone piling up allowable exclusions to such an extraordinary height as he did: of the adult population, some ten-twelfths were excludable (one-half by sex; at least half the rest by age; of the remaining quarter, one-third by property). To say the least, this does give grounds for considering Mill less than a whole-hearted democrat. Why did he do it, and especially why did he admit a property qualification? And why, having done this, did he conclude his argument by reverting to his case for universal franchise, and say that it would not be dangerous because the vast majority of the lower class would always be guided by the middle class?

Mill's allowing such exclusions may be due to the fact that he, like Bentham, was primarily interested in an electoral reform which would undermine the dominant sinister interest of the narrow landed and moneyed class which was in full control

[33] Ibid., p. 49.
[34] Ibid., p. 50.
[35] Ibid., p. 50.

before the 1832 Reform Bill. About this he was much more of an activist than Bentham: he was not above trying, with some success, to frighten the oligarchy into granting the 1832 Reform (which was far short of manhood suffrage), by holding out the likelihood of a popular revolution if such reform were not granted, though it is doubtful if he himself believed in the likelihood of such revolutionary action.[36] But he was very much aware of the importance of getting both working-class and middle-class support for such reform: he was convinced of the importance of public opinion, including the opinion of both those classes. In pressing for reform, therefore, he must avoid offending either class.

Now Mill would not offend either class by permitting the exclusion of women: as Bentham at least believed, probably quite correctly, public opinion was far from ready to admit women to the franchise. The notion of excluding all men under the age of 40 was so palpably absurd that it would not offend anybody. One might indeed argue that such an exclusion would reduce the number of working-class voters more than in proportion to the well-to-do, in view of the smaller proportion of the poor who reached the age of 40, but this point does not seem to have been taken up by Mill's critics: Macaulay, much his most exhaustive critic, did draw attention to the incompetence of Mill's case for excluding women,[37] but made no reference to the case for excluding the under-forties: presumably he thought it beneath notice.

The only difficult decision for Mill was what to say about a property qualification. To advocate full manhood suffrage with no property qualification would frighten much middle-class opinion; to advocate a property qualification which would exclude a substantial part of the working class would be to lose their support. So Mill found himself in a position which is, oddly enough, parallel to that which he attributed to the

[36] Cf. Joseph Hamburger: *James Mill and the Art of Revolution*, New Haven, 1963, especially ch. 3.

[37] Macaulay: 'Mill's Essay on Government', *Edinburgh Review*, March 1829, reprinted in *The Miscellaneous Writings and Speeches of Lord Macaulay*, London, Longmans, Green, 1889 (Popular Edition), p. 174.

spokesmen of what he called the opposition party of the ruling class, and he took the same way out.

In an article in the first number of the radical *Westminster Review* (January 1824) on 'Periodical Literature', Mill launched an abrasive attack on the *Edinburgh Review*, which he said spoke for the anti-Ministerial wing of the ruling class. The dilemma of that party, he said, was that, in order to discredit the Ministry so as to get themselves in, they needed to enlist non-ruling-class opinion, since that opinion did operate upon the ruling class 'partly by contagion, partly by conviction, partly by intimidation'; yet they could not take a position against the present privileges of the ruling class, support from as many as possible of whom they primarily needed to get themselves in, and of which they were of course themselves a part. 'In their speeches and writings, therefore, we commonly find them playing at *seesaw*.' Now they recommend the interests of the ruling class, now the interests of the people. 'Having written a few pages on one side, they must write as many on the other. It matters not how much the one set of principles are really at variance with the other, provided the discordance is not very visible, or not likely to be clearly seen by the party on whom it is wished that the delusion should pass.'[38]

Mill's seesaw in the article *Government* is quite parallel: the discordance between his two sets of principles, the one requiring universal franchise, the other permitting enormous exclusions, is kept 'not very visible' by his recommending a restricted franchise only hypothetically. He later denied that he was *advocating* the exclusion of women, any more than that of men under the age of forty; his son reports him as having said that he was only asking what was the utmost allowable limit of restriction assuming that the franchise was to be restricted;[39] but the wording of the article suggests not that he regarded the restrictions as unfortunately necessary concessions to political realism, but rather that he regarded them as useful in securing that the electors would make a good choice.[40]

[38] *Westminster Review*, i. 218.

[39] J. S. Mill: *Autobiography*, ed. Laski, Oxford World's Classics, 1924, pp. 87–8.

[40] e.g. his statement that 'a very low [property] qualification is of no

The seesaw in the article *Government* is completed by Mill's assurance to his readers, at the very end of the article, that no danger was to be anticipated from any enfranchisement of the lower class because the great majority of that class would always be guided by the middle class. Such reassurance to his middle-class readers Mill might have thought advisable, since even the exclusion of the poorest one-third of the males might be calculated to leave the working class in the majority.

Ten years after the article *Government*, and six years after his analysis of the seesaw, he felt able to make his position somewhat clearer. In an article devoted to advocating the secret ballot, he wrote: 'Our opinion, therefore, is that the business of government is properly the business of the rich, and that they will always obtain it, either by bad means, or good. Upon this every thing depends. If they obtain it by bad means, the government is bad. If they obtain it by good means, the government is sure to be good. The only good means of obtaining it are, the free suffrage of the people.'[41] This catches nicely the best spirit of Model 1, the high point of its optimism: the democratic franchise would not only protect the citizens, but would even improve the performance of the rich as governors. It is scarcely a spirit of equality.

PROTECTIVE DEMOCRACY FOR MARKET MAN

This was the genesis of the first modern model of democracy. It is neither inspiring nor inspired. The democratic franchise provisions were put in the model only belatedly. It is hard to say what had the greater effect in moving the founders of this model to make their franchise democratic in principle: whether it was their realization that nothing less than 'one man, one vote' would placate a working class which was showing signs of becoming seriously politically articulate (as is suggested by Bentham's remark in 1820 that he supposed they wouldn't be satisfied with less), or whether it was the sheer logic of their own case for reform, resting as it did on the assumption of

---

use, as affording no security for a good choice beyond that which would exist if no pecuniary qualification was required' (Barker ed., p. 49).

[41] 'On the Ballot', *Westminster Review*, July 1830.

conflicting self-interested maximizing individuals. Either way, it is clear that they allowed themselves a democratic conclusion only because they had convinced themselves that a vast majority of the working-class would be sure to follow the advice and example of 'that intelligent, that virtuous rank', the middle class. It is on that note that James Mill closed his somewhat ambiguous case for a democratic franchise.

In this founding model of democracy for a modern industrial society, then, there is no enthusiasm for democracy, no idea that it could be a morally transformative force; it is nothing but a logical requirement for the governance of inherently self-interested conflicting individuals who are assumed to be infinite desirers of their own private benefits. Its advocacy is based on the assumption that man is an infinite consumer, that his overriding motivation is to maximize the flow of satisfactions, or utilities, to himself from society, and that a national society is simply a collection of such individuals. Responsible government, even to the extent of responsibility to a democratic electorate, was needed for the protection of individuals and the promotion of the Gross National Product, and for nothing more.

I have drawn a harsh, but I think fair, portrait of the founding model of modern Western democracy. It has nothing in common with any of the earlier, pre-industrial visions of a democratic society. The earlier visions had asked for a new kind of man. The founding model of liberal democracy took man as he was, man as he had been shaped by market society, and assumed that he was unalterable. It was on this point chiefly that John Stuart Mill and his humanist liberal followers in the twentieth century attacked the Benthamist model. But as we shall see, in the next chapter, they were not able to get entirely away from it. For that model did fit, remarkably well, the competitive capitalist market society and the individuals who had been shaped by it. And that society and those individuals were still well entrenched, in spite of the humanist revulsion against them, later in the nineteenth century and in the twentieth. The revulsion was what sparked the formulation of Model 2, first by John Stuart Mill; but the entrenchment of the market society and market man sapped the strength of Model 2 from the beginning.

III

Model 2: Developmental Democracy

THE EMERGENCE OF MODEL 2

We have seen that Bentham and James Mill had no vision of a new kind of society or a new kind of man. They did not need such a vision, because they did not question that their model of society—the hard-driving competitive market society with all its class-division—was justified by its high level of material productivity, and that the inequality was inevitable. In any case, it was a law of human nature that every individual would always be trying to exploit everyone else, so nothing could be done about society. All that could be done was to prevent governments oppressing the governed, and for this a mechanical protective democratic franchise was sufficient.

But by about the middle of the nineteenth century two changes in that society were thrusting themselves on the attention of liberal thinkers, changes which required a quite different model of democracy. One change was that the working class (which Bentham and James Mill had thought not dangerous) was beginning to seem dangerous to property. The other was that the condition of the working class was becoming so blatantly inhuman that sensitive liberals could not accept it as either morally justifiable or economically inevitable. Both these changes raised new difficulties for liberal-democratic theory—difficulties which, as we shall see, were never fully overcome. But those changes did make it clear that a new model of democracy was needed. It was first provided by John Stuart Mill.

That the younger Mill did arrive at his Model 2 because of the two actual changes is evident from his own writings. He was very much aware of the growing militancy of the working class: the revolutions of 1848 in Europe, and the phenomenon of the Chartist movement in England, made a strong impression on him. So did the increasing literacy of the working class, the spread of working-class newspapers, and the increase in working-class organizing ability shown in the growth of trade unions and mutual benefit societies. Mill was convinced that 'the poor' could not be shut out or held down much longer.

Thus in the *Political Economy* he wrote, in 1848:

> Of the working men, at least in the more advanced countries of Europe, it may be pronounced certain, that the patriarchal or paternal system of government is one to which they will not again be subject. That question was decided, when they were taught to read, and allowed access to newspapers and political tracts; when dissenting preachers were suffered to go among them, and appeal to their faculties and feelings in opposition to the creeds professed and countenanced by their superiors; when they were brought together in numbers, to work socially under the same roof; when railways enabled them to shift from place to place, and change their patrons and employers as easily as their coats; when they were encouraged to seek a share in the government, by means of the electoral franchise. The working classes have taken their interests into their own hands, and are perpetually showing that they think the interests of their employers not identical with their own, but opposite to them. Some among the higher classes flatter themselves that these tendencies may be counteracted by moral and religious education: but they have let the time go by for giving an education which can serve their purpose. The principles of the Reformation have reached as low down in society as reading and writing, and the poor will not much longer accept morals and religion of other people's prescribing. . . . The poor have come out of leading-strings and cannot any longer be governed or treated like children. . . . Whatever advice, exhortation or guidance is held out to the labouring classes, must henceforth be tendered to them as equals, and accepted by them with their eyes open. The prospect of the future depends on the degree in which they can be made rational beings.[1]

The conclusion that something must be done had been made explicit in 1845 in the lesson he drew from the Chartist movement.

[1] *Principles of Political Economy*, Bk IV, ch. 7, sects. 1 and 2; in *Collected Works*, ed. J. M. Robson, Toronto and London, 1965, iii. 761–3.

The democratic movement among the operative classes, commonly known as Chartism, was the first open separation of interest, feeling, and opinion, between the labouring portion of the commonwealth and all above them. It was the revolt of nearly all the active talent, and a great part of the physical force, of the working classes, against their whole relation to society. Conscientious and sympathizing minds among the ruling classes, could not but be strongly impressed by such a protest. They could not but ask themselves, with misgiving, what there was to say in reply to it; how the existing social arrangements could best be justified to those who deemed themselves aggrieved by them. It seemed highly desirable that the benefits derived from those arrangements by the poor should be made less questionable—should be such as could not easily be overlooked. If the poor had reason for their complaints, the higher classes had not fulfilled their duties as governors; if they had no reason, neither had those classes fulfilled their duties in allowing them to grow up so ignorant and uncultivated as to be open to these mischievous delusions. While one sort of minds among the more fortunate classes were thus influenced by the political claims put forth by the operatives, there was another description upon whom that phenomenon acted in a different manner, leading, however, to the same result. While some, by the physical and moral circumstances which they saw around them, were made to feel that the condition of the labouring classes *ought* to be attended to, others were made to see that it *would* be attended to, whether they wished to be blind to it or not. The victory of 1832, due to the manifestation, though without the actual employment, of physical force, had taught a lesson to those who, from the nature of the case, have always the physical force on their side; and who only wanted the organization, which they were rapidly acquiring, to convert their physical power into a moral and social one. It was no longer disputable that something must be done to render the multitude more content with the existing state of things.[2]

One of the things that had to be done 'to render the multitude more content with the existing state of things' was to abandon or transform the Benthamite models of man and society. Although John Stuart Mill hoped that the working class might in the future become rational enough to accept the laws of political economy (as he understood them), he could not expect that they would accept Bentham's view that the working class was inevitably doomed to near-indigence. Nor

[2] 'The Claims of Labour' (1845), reprinted in *Dissertations and Discussions* (1867), ii. 188–90; *Collected Works*, ed. Robson, 1967, iv. 369–70.

did he want them to accept that view, which he believed to be false. He thought they could pull themselves up out of their miserable condition. And he was anxious that they should do so, for he was morally revolted by the life they were compelled to lead. The extent of Mill's abandonment or transformation of the Benthamite models of man, of society, and of democracy, will appear as we look closely (in the next section) at Mill's theory, but some of the essential differences can be sketched now.

The striking difference in the models of democracy is in the purpose which a democratic political system was supposed to have. Mill did not overlook the sheerly protective function of a democratic franchise—the function of which James Mill and Bentham had made so much. The people needed to be protected against the government: 'human beings are only secure from evil at the hands of others, in proportion as they have the power of being, and are, self-*protecting*.'[3] But he saw something even more important to be protected, namely, the chances of the improvement of mankind. So his emphasis was not on the mere holding operation, but on what democracy could contribute to human development. Mill's model of democracy is a moral model. What distinguishes it most sharply from Model 1 is that it has a moral vision of the possibility of the improvement of mankind, and of a free and equal society not yet achieved. A democratic political system is valued as a means to that improvement—a necessary though not a sufficient means; and a democratic society is seen as both a result of that improvement and a means to further improvement. The improvement that is expected is an increase in the amount of personal self-development of all the members of the society, or, in John Stuart Mill's phrase, the 'advancement of community . . . in intellect, in virtue, and in practical activity and efficiency'. The case for a democratic political system is that it promotes this advancement better than any other political system as well as making the best use of the amount of 'moral, intellectual and active worth already existing, so as to operate with the greatest effect on public affairs'.[4] The worth of an

[3] *Considerations on Representative Government*, ch. 3, in *Collected Works*, ed. J. M. Robson, vol. xix, Toronto and London, 1977, p. 404.

[4] Ibid., ch. 2, p. 392.

individual is judged by the extent to which he develops his human capacities: 'the end of man . . . is the highest and most harmonious development of his powers to a complete and consistent whole.'[5]

This takes us to the root of Mill's model of democracy. The root is a model of man very different from that on which Model 1 was based. Man is a being capable of developing his powers or capacities. The human essence is to exert and develop them. Man is essentially not a consumer and appropriator (as he was in Model 1) but an exerter and developer and enjoyer of his capacities. The good society is one which permits and encourages everyone to act as exerter, developer, and enjoyer of the exertion and development, of his or her own capacities. So Mill's model of the desirable society was very different from the model of society to which Model 1 of democracy was fitted.

In offering this model of man and of the desirable society Mill set the tone which came to prevail in liberal-democratic theory, and which dominated at least the Anglo-American concept of democracy until about the middle of the twentieth century. The narrowing stipulation John Stuart Mill put in his model was dropped by later advocates of developmental democracy, but the central vision and the argument for it stayed much the same. This is the democracy of L. T. Hobhouse and A. D. Lindsay and Ernest Barker, of Woodrow Wilson and John Dewey and R. M. MacIver: it is the democracy that World War I was to make the world safe for. It still touches a chord, especially when liberal societies are confronted by totalitarian ones, although as we shall see it has now been pretty well rejected in favour of what is said to be a more realistic model, the Model 3 that we shall be examining in the next chapter. But Model 2 is worth considerable attention, if only because efforts now being made to go beyond Model 3, to re-moralize democracy under the banner of participatory democracy (our Model 4), encounter some of the same difficulties as did Model 2, and will need to learn from its failure.

The difficulties encountered by Model 2 in its first formulation

[5] *On Liberty*, ch. 3; in *Collected Works*, xviii. 261, quoting Humboldt.

were somewhat different from those that beset the later version. So it will be useful to look at the two versions in turn, as Models 2A and 2B. One difference between them may be stated briefly in advance. Mill had been deeply troubled by the incompatibility he saw between the claims of equal human development and the existing class inequalities of power and wealth. Although he did not identify the problem accurately, and so was unable to resolve it even in theory, he did see that there was a problem and did try to deal with it, at least to the extent of concerning himself with the necessary social and economic prerequisites of democracy. His twentieth-century followers scarcely saw this as a problem, at least not as the central problem: when they did not let it drop virtually out of sight, they treated it as something which would or could be overcome in one way or another—for instance, by a revival of idealist morality, or a new level of social knowledge and communication.

Indeed one can see a cumulative decline in realism from Model 1 through Models 2A and 2B. Bentham and James Mill, in formulating Model 1, had recognized that capitalism entailed great class inequalities of power and wealth: they were realistic about the necessary structure of capitalist society, though untroubled by it since it did not conflict with their merely protective democracy. John Stuart Mill, in his Model 2A, was less realistic about the necessary structure of capitalist society: he saw the *existing* class inequality, and saw that it was incompatible with his developmental democracy, but thought it accidental and remediable. The twentieth-century exponents of developmental democracy (our Model 2B) were even less realistic than Mill on this score: they generally wrote as if class issues had given way, or were giving way, to pluralistic differences which were not only more manageable but also positively beneficial. And on top of this there was a new unrealism in Model 2B, a descriptive unrealism.

There had been no question of the two earlier models (1 and 2A) being realistic as *descriptions* of an existing democratic system, for in no country in the nineteenth century were governments chosen by manhood suffrage, let alone universal suffrage.[6] The two earlier models were statements of what would be necessary to achieve at least protection and at best self-development for all. But by the first half of the twentieth century, with at least full manhood suffrage the general rule in advanced Western countries, a model could reasonably be expected also to be realistic as a descriptive statement. Model 2B did offer itself as a statement of what the existing system essentially *was* (which often meant, rather, what the present imperfect system was capable of becoming), as well as a statement of its desirability. But as a statement of how the democratic system actually worked Model 2B was seriously inaccurate, as was demonstrated by the exponents of Model 3. Model 2B may thus be said to have been doubly unrealistic: it failed both to grasp the necessary implications of capitalist society and to describe the actual twentieth-century liberal-democratic system.

To anticipate our argument one further stage, it may now be said that the currently prevalent Model 3, which boasts its realism both as a descriptive and explanatory model and as a demonstration of the necessary limits of the democratic principle of effective citizen participation, will be found to fall short on both counts.

MODEL 2A: J. S. MILL'S DEVELOPMENTAL DEMOCRACY

I have emphasized how different J. S. Mill's model of a desirable society was from Bentham's and James Mill's. The difference can be made more precise. Bentham and James Mill accepted existing capitalist society without reservation; John Stuart Mill did not. The difference is clearly expressed in the

[6] Although most states in the United States had manhood white franchise by about the middle of the nineteenth century, manhood franchise can scarcely be said to have been effectively in existence in the United States until the twentieth century. A few European countries in the nineteenth century (France 1848, Germany 1871) had manhood franchise for the national assembly, but the assembly did not choose or control the government. In the United Kingdom, as late as 1911 only 59 per cent of adult males had the franchise, that is, had their names on the parliamentary electoral roll. See Neal Blewett: 'The Franchise in the United Kingdom 1885–1918', *Past and Present*, no. 32 (Dec. 1965).

younger Mill's position on the desirability of 'the stationary state' which he, like they, thought would be the culmination of capitalism: they regarded it with dismay, he welcomed it. As he put it in 1848:

> I confess I am not charmed with the ideal of life held out by those who think that the normal state of human beings is that of struggling to get on; that the trampling, crushing, elbowing, and treading on each other's heels, which form the existing type of social life, are the most desirable lot of human kind, or anything but the disagreeable symptoms of one of the phases of industrial progress. It may be a necessary stage in the progress of civilization . . . But it is not a kind of social perfection which philanthropists to come will feel any very eager desire to assist in realizing . . . In the meantime, those who do not accept the present very early stage of human improvement as its ultimate type, may be excused for being comparatively indifferent to the kind of economic progress which excites the congratulations of ordinary politicians; the mere increase of production and accumulation.[7]

Society, in the vision of Model 2, need not be, should not be, what Model 1 had assumed it was and always would be. It need not be and should not be a collection of competing, conflicting, self-interested consumers and appropriators. It could and should be a community of exerters and developers of their human capacities. But it was not that now. The problem was to get it to advance to that. The case for democracy was that it gave all the citizens a direct interest in the actions of the government, and an incentive to participate actively, at least to the extent of voting for or against the government, and, it was hoped, also of informing themselves and forming their views in discussions with others. Compared with any oligarchic system, however benevolent, democracy drew the people into the operations of government by giving them all a practical interest, an interest which could be effective because their votes could bring down a government. Democracy would thus make people more active, more energetic; it would advance them 'in intellect, in virtue, and in practical activity and efficiency'.

[7] *Principles of Political Economy*, Bk. IV, ch. 6, sect. 2; in *Collected Works*, iii. 754–5.

This is a rather large claim to make for a system of representative government in which the ordinary person's political activity is confined to voting every few years for a member of Parliament, perhaps a little oftener for local councillors, and perhaps actually holding some elective local office. Even so, the claim might be allowed by contrast with any oligarchic system, which positively discourages general interest and involvement. By that contrast, democracy might seem to lead to self-sustaining, even self-increasing, advancement of the citizens in moral, intellectual, and active worth, every bit of participation giving an ability and an appetite for more.

But here Mill came up against a difficulty which turned out to be insuperable. To see what it was we must look at another basic difference between John Stuart Mill and Bentham. Underlying the difference in their moral evaluations of existing society was a difference in their definitions of happiness or pleasure, the thing they both held should be maximized.

Bentham had held that in calculating the greatest happiness one need take into account only the amounts of undifferentiated pleasure (and pain) actually felt by the individuals. There were no qualitative differences between pleasures: pushpin was as good as poetry. And since, as we have seen, he measured pleasure or utility in terms of material wealth, the aggregate greatest happiness of the whole society was to be attained by maximizing productivity (though even that conclusion was fallacious, as we have noticed).

J. S. Mill insisted, on the contrary, that there were qualitative differences in pleasures, and he refused to equate the greatest aggregate happiness with maximum productivity. The greatest aggregate happiness was to be got by permitting and encouraging individuals to develop themselves. That would make them capable of higher pleasures, and so would increase the aggregate pleasure measured in both quantity and quality.

But at the same time—and this was the fundamental difficulty—Mill recognized that the existing distribution of wealth and of economic power made it impossible for most members of the working class to develop themselves at all, or even to live humanly. He denounced as utterly unjust that the produce of labour should be apportioned as we now see it, almost in an inverse ratio to the labour—the largest portions to those who have never worked at all, the next largest to those whose work is almost nominal, and so in a descending scale, the remuneration dwindling as the work grows harder and more disagreeable, until the most fatiguing and exhausting bodily labour cannot count with certainty on being able to earn even the necessaries of life . . .[8]

This, he said, was the very opposite of the only 'equitable principle' of property, the principle of 'proportion between remuneration and exertion'. That *was* the equitable principle because the only justification of the institution of private property was that it guaranteed to individuals 'the fruits of their own labour and abstinence', not 'the fruits of the labour ond abstinence of others'.[9]

A few pages later Mill gave an extended definition of property:

The institution of property, when limited to its essential elements, consists in the recognition, in each person, of a right to the exclusive disposal of what he or she have produced by their own exertions, or received either by gift or by fair agreement, without force or fraud, from those who produced it. The foundation of the whole is, the right of producers to what they themselves have produced.[10]

This seems a reasonable extension of the principle first announced, at least as far as 'fair agreement' is concerned, though 'gift' raises a problem. Without a property right in what one has exchanged by agreement for the fruits of one's labour, not even the simplest exchange economy would be possible. But Mill is talking about a capitalist exchange economy, where the produce is the result of the combination of current labour with capital provided by someone else, and where the labourer gets as his share only a wage, and the capitalist gets the rest, both shares being determined by market competition. Mill held that this relation was justified also. Speaking of the capitalist's acquisition from the wage contract, he wrote:

The right of property includes, then, the freedom of acquiring by contract. The right of each to what he has produced, implies a right

[8] Ibid., Bk. II, ch. 1, sect. 3, p. 207.

[9] Ibid., p. 208.

[10] Ibid., Bk. II, ch. 2, sect. 1, p. 215.

to what has been produced by others, if obtained by their free consent; since the producers must either have given it from good will, or exchanged it for what they esteemed an equivalent, and to prevent them from doing so would be to infringe their right of property in the product of their own industry.[11]

The owner of the capital, Mill saw, must have a share of the product, and he held that this was consistent with the equitable principle because capital is simply the product of previous labour and abstinence. This justified the distribution of the product between wage-labourers and owners of capital: given competition between capitalists for labourers, and between labourers for employment, there was a fair division between those who contributed current labour and those who contributed the fruits of past labour and abstinence. Mill acknowledged that the capital was not usually created by the labour and abstinence of the present possessor, but thought he had made a sufficient case for the labour/capital distribution by saying that the present possessor of capital 'much more probably' got it by gift or voluntary contract than by wrongful dispossession of those who had created it by their past labour.[12]

The fact that the present possessors may have got some of their capital by gift, i.e. by inheritance, gave Mill some uneasiness: it seemed clearly inconsistent with his equitable principle of property. But he held that the right to dispose of one's property by bequest was an essential part of the right of property. The farthest he was willing to go was to recommend a limit on the amount any one person could inherit, but he set the limit so high—each could inherit enough 'to afford the means of comfortable independence'[13]—that this did nothing to resolve the inconsistency. Mill fell back on the argument that 'while it is true that the labourers are at a disadvantage compared with those whose predecessors had saved, it is also true that the labourers are far better off than if those predecessors had not saved.'[14]

[11] Ibid., p. 217.
[12] Ibid., pp. 215–16.
[13] Ibid., Bk. II, ch. 2, sect. 4, p. 225.
[14] Ibid., Bk. II, ch. 2, sect. 1, p. 216.

Thus Mill was satisfied that there was no inconsistency between his equitable principle of property—reward in proportion to exertion—and the principle of reward in proportion to the market value of both the capital and the current labour required for capitalist production.

Yet, as we have seen, he found the actual prevailing distribution of the produce of labour wholly unjust. He found the explanation of that unjust distribution in an historical accident, not in the capitalist principle itself.

> The principle of private property has never yet had a fair trial in any country; and less so, perhaps, in this country than in some others. The social arrangements of modern Europe commenced from a distribution of property which was the result, not of just partition, or acquisition by industry, but of conquest and violence: and notwithstanding what industry has been doing for many centuries to modify the work of force, the system still retains many and large traces of its origin.[15]

It was this original violent distribution of property, not anything in the principle of private property and capitalist enterprise as such, that had led to the present miserable position of the bulk of the working class, about the injustice of which Mill was so outspoken: 'The generality of labourers in this and most other countries, have as little choice of occupation or freedom of locomotion, are practically as dependent on fixed rules and on the will of others, as they could be on any system short of actual slavery.'[16]

In thus putting the blame on the original feudal forcible distribution of property, and the failure of subsequent property law to rectify it, Mill was able to think that the capitalist principle was not in any way responsible for the existing inequitable distributions of wealth, income, and power, and even to think that it was gradually reducing them. What he failed to see was that the capitalist market relation enhances or replaces any original inequitable distribution, in that it gives to capital part of the value added by current labour, thus steadily increasing the mass of capital. Had Mill seen this he could not

[15] Ibid., Bk. II, ch. 1, sect. 3, p. 207.
[16] Ibid., p. 209.

have judged the capitalist principle consistent with his equitable principle. Failing to see this, he found no fundamental inconsistency, and was not troubled by it.

However, the present debased position of the bulk of the working class did present an immediate and serious problem to Mill, and he met it forthrightly. The difficulty was that in their present condition they were incapable of using political power wisely. Mill believed indeed that people were *capable* of becoming something other than self-interested acquirers of benefit for themselves, but he thought that most of them had not yet got much beyond that. It would be foolish, he said, to expect the average man, if given the power to vote, to use it with 'disinterested regard for others, and especially for what comes after them, for the idea of posterity, of their country, or of mankind'.

> Governments must be made for human beings as they are, or as they are capable of speedily becoming: and in any state of cultivation which mankind, or any class among them, have yet attained, or are likely soon to attain, the interests by which they will be led, when they are thinking only of self-interest, will be almost exclusively those which are obvious at first sight, and which operate on their present condition.[17]

This being so, what would happen if everyone had a vote? Presumably the selfish society would continue.

But there was worse to be feared than that. For Mill recognized that modern societies were divided into two classes with interests which they believed to be opposed, and which in important respects Mill granted were opposed. The classes were, roughly, the working class (in which he included petty tradesmen) and the employing class, including those who lived on unearned income and those 'whose education and way of life assimilate them with the rich'.[18] The working class was of course the more numerous. 'One person, one vote' would therefore mean class legislation in the supposed immediate interest of one class, who must be expected 'to follow their own selfish inclinations and short-sighted notions of their own good, in

[17] *Representative Government*, ch. 6; in *Collected Works*, xix. 445.
[18] Ibid., p. 447.

opposition to justice, at the expense of all other classes and of posterity'.[19] Something must therefore be done to prevent the more numerous class from being able to 'direct the course of legislation and administration by its exclusive class interest' (even though this would be less of an evil than the present class rule by a small class based merely on established wealth).[20]

Mill's dilemma was a real one, for his main case for a universal franchise was that it was essential as a means of getting people to develop themselves by participation. Mill's way out was to recommend a system of plural voting for members of the smaller class, such that neither of the two classes should outweigh the other, and neither therefore would be able to impose 'class legislation'.[21]

Everyone should have a vote, but some should have several votes. Or rather, everyone with certain exceptions should have a vote, and some should have several votes. In his *Thoughts on Parliamentary Reform*, published in 1859, Mill held that a perfect electoral system required both that every person should have one vote and that some should have more than one vote, and said that neither of these provisions was admissible without the other. But in *Representative Government* (1861) he argued for plural votes for some along with the exclusion of others from any vote at all. The exclusions reflect Mill's acceptance of the standards of the market society. Those in receipt of poor relief were to be excluded: they had failed in the market. So were undischarged bankrupts. So were all who did not pay direct taxes. Mill knew that the poor paid indirect taxes, but, he said, they didn't feel them, and therefore would be reckless in using their votes to demand government largess. The direct tax requirement was not intended to deprive the poor of a vote: the way out was to replace some of the indirect taxes by a direct head tax which even the poorest would pay. Again, those who could not read, write, and reckon, were to be excluded. This also was not intended as a back-handed way of excluding a large number of the poor, for Mill held that society had a duty to put elementary schooling within reach of

[19] Ibid., p. 446.
[20] Ibid., ch. 8, p. 467.
[21] Ibid., ch. 8, p. 476.

all who wanted it. But it *would* effectively have excluded the poor, for he held that when society had failed to perform this duty (as it clearly had in Mill's time), the exclusion from the franchise of those who suffered from that failure was 'a hardship that ought to be borne'.[22]

Whether or not any of these provisions would have excluded a significant number of the working class, plural voting was still needed, and was recommended on an additional ground. The system of plural voting would not only prevent class legislation: it would be positively beneficial by giving more votes to 'those whose opinion is entitled to a greater weight'[23] by virtue of their superior intelligence, or the superior development of their intellectual or practical abilities. The rough test of this was the nature of a person's occupation: employers, men of business, and professional people are by the nature of their work generally more intelligent or more knowledgeable than ordinary wage-earners, so they should have more votes. Foremen, as more intelligent than ordinary labourers, and skilled labourers as more intelligent than unskilled, might also be allowed more than one vote each. To meet Mill's stipulation that the working class as a whole should not have more votes than the employing and propertied class, members of the latter would have to be given considerably more than two votes each, but Mill excused himself from working out the details. The closest he came to doing so was his suggestion in *Thoughts on Parliamentary Reform* that, if the unskilled labourer had one vote, a skilled labourer should have two; a foreman perhaps three; a farmer, manufacturer, or trader, three or four; a professional or literary man, an artist, a public functionary, a university graduate, and an elected member of a learned society, five or six.[24] Mill's gradations are revealing: the entrepreneur ('farmer, manufacturer, or trader'), with three or four votes, is not much preferred to the foreman, while the intellectuals, artists, and professional people, with five or six votes, are the strongly preferred rank. It is curious,

[22] Ibid., ch. 8, p. 470.
[23] Ibid., ch. 8, p. 474.
[24] *Collected Works*, xix. 324-5.

incidentally, in view of Mill's concern for the rights of women, that he did not suggest how the entitlement of those women who were neither employed nor employers, nor professional or propertied persons, to plural votes was to be determined.

The important point of principle in all this is that Mill argued explicitly that plural voting on grounds of superior attainments was *positively* desirable, not merely negatively desirable as a way of preventing class legislation:

> I do not propose the plurality as a thing in itself undesirable, which, like the exclusion of part of the community from the suffrage, may be temporarily tolerated while necessary to prevent greater evils. I do not look upon equal voting as among the things which are good in themselves, provided they can be guarded against inconveniences. I look upon it as only relatively good; less objectionable than inequality of privilege grounded on irrelevant or adventitious circumstances, but *in principle wrong*, because recognizing a wrong standard, and exercising a bad influence on the voter's mind. It is not useful, but hurtful, that the constitution of the country should declare ignorance to be entitled to as much political power as knowledge.[25]

So John Stuart Mill cannot be ranked as a full egalitarian. Some individuals were not only better than others, but better in ways directly relevant to the political process, better in ways that entitled them to more political weight. True, part of the reason why they were to be given greater weight was that this would make for a better society, at least negatively: it would reduce the likelihood of short-run narrowly selfish interests being predominant in legislation and government, which would be the outcome of equal weighting. Unequal weighting would be more likely to lead to a society democratic in the best sense, a society where everyone could develop his or her human capacities to the fullest. Nevertheless, unequal political weights for citizens were built into Mill's model on a ground which seems more permanent: as long as people were unequal in knowledge (and when would they not be?) equal weighting was wrong in principle.

The weighting Mill gave to knowledge and skill led him also to recommend that Parliament should not itself initiate any

[25] *Representative Government*, ch. 8, p. 478 (my italics).

legislation but should be confined to approving or rejecting, or sending back for reconsideration but not itself amending, legislative proposals all of which would be sent up to it by an expert non-elected Commission. Mill's impatience with existing parliamentary and cabinet procedure is understandable, but his remedy would reduce the power of the elected legislature, and so would contribute to the disincentive of democratic voters to participate in the electoral process. If he realized this, he didn't mind it, such was the premium he placed on expertise.

So Mill's model, the original version of Model 2, is arithmetically a step backward from Model 1, which had stipulated, in principle at least, 'one person, one vote'. But in its moral dimension Model 2 is more democratic than Model 1. Model 2 is not satisfied with individuals as they are, with man as infinite consumer and appropriator. It wants to move towards a society of individuals more humanly developed and more equally so. It wants not to impose a utopia on the people but to have the people reach the goal themselves, improving themselves by participating actively in the political process, every instalment of participation leading to an improvement in their political capacity, as well as their all-round development, and making them capable of more participation and more self-development.

It is easy now to point to defects and contradictions in Mill's model. An obvious one is in the matter of participation and self-development. Participation in the political process was necessary to improve people's quality and would improve it. But participation with equal weight now would reinforce low quality. Therefore those who had already attained superior quality, as judged by their education or station in life, must not be made to yield their power to the rest. In the name of equal self-development, a veto is given to those who are already more developed. But the less developed individuals within Mill's model, if they stayed within it (that is, if they accepted the inferior electoral weight Mill gave them), would know that their wills could not prevail, so would not have much incentive to participate, so would not become more developed.

A deeper difficulty, which is at the root of that one, is in

Mill's model of man and of society. Men as shaped by the existing competitive market society were not good enough to make themselves better. Mill deplored the effects of the existing market society on the human character, which made everybody an aggressive scrambler for his own material benefit. He deplored most strongly the existing relation between capital and labour, which debased both capitalist and labourer. He believed there could not be a decently human society until that relation was transformed. He put his hopes on an enormous spreading of producers' co-operatives, whereby workmen would become their own capitalists and work for themselves jointly. He allowed himself to hope that producers' co-ops would call forth such better workmanship, and thus be so much more efficient units of production, that they would displace the capitalist organization of production.

Yet he accepted and supported the received capitalist property institutions, at least until such time as they had been modified or transformed by his producer's co-ops; and even then the competitive market system would still operate, for the separate co-operative enterprises were expected to compete in the market, and would be driven by the incentive of desire for individual gain. In other words, Mill accepted and supported a system which required individuals to act as maximizing consumers and appropriators, seeking to accumulate the means to ensure their future flow of consumer satisfactions, which meant seeking to acquire property. A system which requires men to see themselves, and to act, as consumers and appropriators, gives little scope for most of them to see themselves and act as exerters and developers of their capacities. Mill did indeed hold out the prospect that the spread of co-operatives would bring a 'moral revolution to society':

> the healing of the standing feud between capital and labour; the transformation of human life, from a conflict of classes struggling for opposite interests, to a friendly rivalry in the pursuit of a good common to all; the elevation of the dignity of labour; a new sense of security and independence in the labouring class; and the conversion of each human being's daily occupation into a school of the social sympathies and the practical intelligence.[26]

[26] *Political Economy*, Bk. IV, ch. 7, sect. 6; in *Collected Works*, iii. 792. This

These high hopes remained unfulfilled. Class opposition continued, and so long as it was not offset in other ways it would still require Mill's watering down of democracy. For the rational behaviour of each of those classes is to try to overbear the opposed class, hence the danger Mill saw of class government, hence the need to deny as much political weight to each member of the more numerous class as to each member of the less numerous class, hence the vicious circle of unequal participation justifying continued unequal participation.

The failure of the co-operative solution thus left unresolved the contradiction Mill saw between a universal equal franchise and the greatest happiness of society. There was no way out, given his assumption that the working class would use an equal franchise to enact class legislation not consistent with the long-run, qualitative, greatest happiness of the whole society.

And underlying that contradiction was the other one, the contradiction between capitalist relations of production as such and the democratic ideal of equal possibility of individual self-development. This contradiction Mill never fully saw. He came close to seeing it in his strictures on the *existing* labour–capital relation (especially when he was contrasting it morally with the co-operative relation); but, as we have noticed, in his analysis of capitalist market relations as such, he justified private property in capital, and the wage-contract, as being consistent in principle with an equitable system.

One might think that the existence of two such serious shortcomings in Mill's liberal-democratic theory would have been enough to prevent it maintaining, in the late nineteenth and twentieth centuries, the position it had won in mid-nineteenth century as *the* model of liberal democracy. But this is not quite what happened. And it is easy to see why.

In the first place, the underlying contradiction could be expected to lead to the abandonment of the theory only if Mill's followers had seen it as a flaw in the theory. But in fact,

---

contrasts oddly with Mill's statement in 1838: 'The numerical majority of any society whatever, must consist of persons all standing in the same social position, and having in the main, the same pursuits, namely, unskilled manual labourers . . .' (Bentham', in *Essays on Ethics, Religion and Society, Collected Works,* x. 107).

as we shall see in the last section of this chapter, the later liberal-democratic theorists showed even less recognition than Mill of any fundamental incompatibility between capitalist market relations and the equal possibility of individual self-development. So they could, and did, still hold to Mill's case for developmental democracy.

In the second place, the incompatibility Mill had seen between a universal equal franchise and the existing opposition of class interests seemed, by the beginning of the twentieth century, to have disappeared. Mill's fear of class government if there were a universal equal franchise had turned out to be unfounded, at least for the time being. Bentham and James Mill had been right about the working class following the lead of the middle class, although as I shall suggest they were right for the wrong reasons. In any case, when the first large instalment of manhood equal suffrage was introduced in England in 1884, eleven years after Mill's death, and further instalments later, they did not bring class rule by the working class. So Mill's followers could, and did, cheerfully abandon the inegalitarian provisions of his model—the plural voting and the downgrading of the elected legislature in favour of an expert legislative commission—while holding to his main developmental case.

We should not, therefore, speak of Model 2A as a failure. Its main lines continued to be generally accepted by liberal-democrats, the more easily because its inegalitarian stipulations could be dropped. They were dropped, partly because they came to appear unnecessary, and partly because it became clear that anything of that sort would be unacceptable to forbiddingly strong popular movements.[27] But this enabled the rest of Model 2A to live on, as 2B, well into the twentieth century. The consistent success of the reigning politicians in the nineteenth century, and of the system itself in the twentieth

[27] The strength of such movements was evident in the agitation for the 1867 Reform Bill, of which Mill was a close and concerned observer. He withdrew his undertaking to endorse the radical Reform League when he found that it was appealing to physical force to attain its uncompromising franchise demands. (Mill to W. R. Cremer, 1 March 1867, *Later Letters*; in *Collected Works*, xvi. 1247–8.) See also Royden Harrison: *Before the Socialists, Studies in Labour and Politics 1861–1881*, London and Toronto, 1965, ch. 3.

century, in deflecting the menacing implications of the democratic franchise, delayed the failure of Model 2 until the mid-twentieth century. And it failed then not because its mid-twentieth-century critics, the exponents of Model 3, had realized or exposed the internal contradictions in Model 2, for they did not. It failed for different reasons, which we must now explore.

THE TAMING OF THE DEMOCRATIC FRANCHISE

Before we look at the fortunes of the later developmental model, we must examine the reason why the equal manhood franchise did not bring about the class government that Mill had feared, so that the way was left open for the later liberal-democrats to redeploy Mill's general case. This will help us to understand both the sway of the later developmental model down to about the middle of the twentieth century, and *its* ultimate failure.

What happened was something which Mill did not foresee, perhaps could scarcely have foreseen. But the interesting thing is that the later developmental theorists, those who promoted Model 2B, did not seem to see it or understand it, though they should have been able to see it by then. And I shall suggest that their failure to see it was what led to the failure of 2B and its supersession by Model 3.

The reason that the equal manhood franchise did not bring about the class government Mill had feared was the extraordinary success with which the party system was able to tame the democracy. This is important because, although it gave Model 2 a new lease on life, it was in the end Model 2's undoing. For it left the actual democratic political process largely unable to provide the effective degree of participation its advocates claimed or hoped for it, and unable to promote that personal development and moral community which was the main rationale offered for liberal democracy. It is this which so undermined Model 2 that it could be swept aside in mid-twentieth century by the apparently more realistic Model 3 examined in the next chapter.

How did the party system rescue the developmental model

and enable it to hold the field, in its revised equal-franchise form, for another half century or more? How was the party system able to prevent the class take-over that Mill had feared, and so allow the developmental image of democracy to be maintained by liberal advocates after the equal franchise had been introduced? A universal equal franchise would obviously give the preponderant voice to the wage-earning working class in the more industrialized countries, and to the farmers and other small independent operators (or a mixture of them and wage-earners) in the less industrialized ones, and in both cases a conflict of interests with established capitalist property was to be expected. How could a thing as mechanical and neutral as a system of competing parties prevent the take-over of power by the subordinate but more numerous class or classes? Would not a party system, in so far as it efficiently represented the numerical weight of the different interests, actually bring about the take-over rather than prevent it? Yet the take-over has been prevented, and through the instrumentality of the party system, in all the Western democracies.

The way this has happened has been somewhat different in different countries, depending partly on the class composition of the country, partly on whether there was a responsible non-democratic party system in operation before the arrival of the democratic franchise, and partly on other differences of national traditions. I cannot attempt here an analysis of all the complex differences between the ways the party systems performed the same basic function in countries as different as England, the United States, Canada, and the various Western European nations. Yet it is not difficult to see, if one shifts the focus slightly from that of the usual descriptions of the function of the party system, that its main function is not merely to produce a stable political equilibrium but to produce a particular kind of equilibrium.

I think it is not overstating the case to say that the chief function the party system has actually performed in Western democracies since the inception of a democratic franchise has been to blunt the edge of apprehended or probable class conflict, or, if you like, to moderate and smooth over a conflict of class interests so as to save the existing property institutions and the market system from effective attack. This is less evident in America than in Europe, where the relation between party and class is generally more obvious. And it is less evident than it might be to twentieth-century observers anywhere, because of the very success of the party system in thrusting out of sight class issues which in the nineteenth century had bulked much larger.

The function of blurring class lines and so mediating between conflicting class interests can be seen to be equally well performed by any of three varieties of party system: (1) a two-party (or two dominant parties) system, even where the parties were intended to represent two opposed class interests, as in England with the Labour and Conservative parties; (2) a two-party (or two dominant parties) system where each main party is a loose organization of many regional and sectional interests, as in the United States and Canada; or (3) a multi-party system with so many parties that the government generally has to be a coalition, as in most Western European countries. In the first case, each party tends to move towards a middle position, which requires that it avoid an apparently class position. It must do this in order to be able to project an image of itself as a national party standing for the common good, without which image it fears it will not stand much chance of long-run majority support. In the second case each of the main parties is compelled to act in a similar way, only more so: each must offer a platform which is all things to all men and which is therefore very indefinite. True, in such a system, a third or fourth party may start with a position which has a specific class content, but if such a party grows to a size that puts it within reach of being the second or first party, it has to do the same. In the third case, a really multi-party system, where no one party can usually expect a majority, no party can give an unequivocal undertaking to the electorate because both the party and the electorate know that the party will have to compromise continually in the coalition government.

Now it is true that none of these three blurring systems could have operated as they have done if a bi-polar class-division in the country as a whole had overridden both the sense of national identity and all sectional, religious, ethnic, and other cross-currents. None of the three systems could operate as they do if the numerically largest economic class were a single-minded class, whose members were not pulled in other directions by such cross-currents or by traditional attachments. But as it happened, in all these countries, at the same time the democratic franchise was becoming operative, there were factors which weakened the expected bi-polar division between those who supported and those who seemed likely to reject the existing system of property and of market competition. In nineteenth-century North America, continental expansion and free land made the largest class, independent farmers and other small working proprietors, the epitome of the petty-bourgeoisie: they wanted private capitalism and the market economy, provided only it was not rigged in favour of the capitalists of the commercial metropolises. In the same period, the late nineteenth and early twentieth centuries, the imperial expansion in which England and most of the Western European countries were indulging allowed their governments to afford handouts to their electorates which reduced the working-class pressures for fundamental reforms. Had it not been for these factors, the apparently neutral party system could not have done the job. But given these factors, without the party system it is unlikely that the job could have been done. The party system, in whichever of its variants, was the means by which the job of blurring the still underlying class differences was done.

The party system had a built-in ability to do this because of another feature. With every extension of the franchise, a party system becomes necessarily less responsible to the electorate. Take the classic case of the English party system. It had been the effective means of making and unmaking governments for half a century or more before there was anything like a democratic franchise. As long as the franchise was confined to the propertied class, the relatively small number of electors in each constituency made it possible for the electors to exert considerable influence, even control, over their elected member. And because the M.P.s could thus be held responsible to their constituents, or at least to the active party people in the constituency, i.e. to the constituency party, however loosely organized it might be, they could not be dominated by the cabinet, i.e. the leading men in the parliamentary party.

All this changed with the democratization of the franchise. Appeal to a mass electorate required the formation of well-organized national parties outside the parliamentary parties. Effective organization required centrally controlled party machines. Endorsement by the party machine became virtually the only way of getting elected to Parliament. The central party leadership was therefore able to control its M.P.s. The main power fell to the party leaders in Parliament, for they, i.e. the Prime Minister and his leading cabinet ministers, commanded the threat of expulsion from the party and the threat of dissolving Parliament prematurely, thus compelling new elections. The cabinet was thus enabled to dominate Parliament to a high degree. It still does so.

Not only is it able to do so: it is now required to do so. For the universal franchise brought a change in the basic job the political system had to do, a change which necessitated government control, rather than constituency or outside party control, of the parliamentary party. Before the franchise became democratic, the function of the system was to respond to the needs of shifting combinations of various elements of the propertied class, which could best be done by governments which were responsible, through the M.P.s, to the leading constituents. But with the democratic franchise, the system has had to mediate between the demands of two classes, those with and those without substantial property. This has meant that the system has continually to be arranging compromises, or at least apparent compromises. Continual compromise requires room for manœuvre. It is the government that must have this room. In a multi-party system, where every government is a coalition, this is understood. It is not always understood that room for manœuvre is just as necessary in a two-party (or two major parties) system, where the government is normally all from one party. But room for manœuvre is equally necessary there, for what requires continual compromise is the opposition of interests *in the country*, whether or not that opposition is represented within the government. A government, especially a majority government, cannot have this room for manœuvre if it is held closely responsible even to the parliamentary party, let alone to the outside party as a whole through an annual party convention, or to the constituency parties. Every attempt, by democratic reform parties and movements in parliamentary countries, to make the government and the members of parliament strictly responsible to the popular organization outside has failed. A sufficient reason for the failure is that such strict responsibility does not allow the room for manœuvre and compromise which a government made up entirely from one party must have in order to carry out its function of mediating between opposed class interests in the whole society.

The general conclusion from this glance at the party system is that the party system has been the means of reconciling universal equal franchise with the maintenance of an unequal society. It has done so by blurring the issues and by diminishing government's responsibility to electorates. It has had to do both these things in order to perform the functions required of it in an unequal society. It has thus necessarily failed to induce the widespread popular participation in the political process which Model 2 required, and hence has failed to develop the active individual as citizen, and to promote moral community, as Model 2 expected.

MODEL 2B: TWENTIETH-CENTURY DEVELOPMENTAL DEMOCRACY

While all this was happening, the rationale put forward by liberal democrats remained the developmental case—substantially Mill's case minus the plural voting proposal.

I shall not take time to examine the democratic theories of the early twentieth-century writers in detail. But it may confidently be said that the tone, the ideal, and the basic justification are much the same as Mill's in all the leading English and American theorists of the first half of the twentieth century, whether in the philosophic idealist tradition (Barker, Lindsay, MacIver), or the pragmatist (Dewey), or the modified utilitarian (Hobhouse). The only exceptions were the few theorists who explicitly tried to combine liberal values with some kind of socialism (Cole, Laski), but they did not significantly deflect the liberal tradition. And in the main liberal tradition of that period there was, by comparison even with Mill, a steady decline in the realism of the analyses of liberal democracy.

Mill had seen the contradiction between his developmental ideal and the class-divided and exploitive society of his own time. He failed to resolve it, even in theory, because he had not identified it accurately: he did not see that it was a contradiction between capitalist relations of production as such and the developmental ideal. But at least he did not assume that the democratic political process could itself overcome the class division and exploitation. He put his hopes in other things as well—producers' co-operatives, working-class education, etc. These hopes were not fulfilled, but at least he did not put all the burden on the democratic process itself.

The theorists of the first half of the twentieth century increasingly lost sight of class and exploitation. They generally wrote as if democracy itself, at least a democracy that embraced the regulatory and welfare state, could do most of what could be done, and most of what needed to be done, to bring a good society. They were, indeed, not insensitive to problems of the concentration of private economic power; and they were not friendly towards the individualist ideology, which they saw underlying the existing order. Lindsay, for instance, was strongly against 'the atomic individualism which has dogged modern democratic theory from the beginning', which, oddly, he identified not only with Bentham but also with Marx. And he did not completely accept the existing control of production by capital: 'the application to the government of industry of . . . democratic principles' would be 'the fulfilment' of democracy. But what he thought sufficient for the democratic control of business was some control of monopolistic business. The consumers' sovereignty of a fully competitive market economy was perfectly acceptable. There was nothing wrong with capitalist relations of production as such. In the end, his hope for democracy came down to a more lively flourishing of pluralistic non-political democratic associations 'like churches and universities'.[28]

This neo-idealist pluralism was a strong current in early

[28] A. D. Lindsay: *The Essentials of Democracy*, 2nd edn., London, 1935, pp. 6, 5, 64 ff., 73–4.

twentieth-century liberal-democratic theory. And there was some excuse, or at least some reason, for those theorists' neglect of class division. The democratic party system had apparently solved the problem: it had overcome the danger of class government. But they did not see *how* it had done this, that is, by reducing the democratic responsiveness of governments to electorates, and so preventing class division from operating politically in any effective way. So they could, and did, write as if the democratic process were an arrangement whereby rational, well-intentioned citizens, who had of course a whole variety of different interests, could adequately adjust their differences in the peaceful, rational, give-and-take of parties and pressure groups and the free press. They allowed themselves to hope that the class issue would go away: either that it was already being replaced by pluralistic social groups, or that it would be so reduced by the welfare and regulatory state that a democratic society would be consistent with a capitalist market society.

Thus Barker, while seeing an amount of 'class-debate' that required giving some attention to 'reckoning gain and loss between different classes and sections', and while recognizing that some redistribution of rights between classes might be necessary if 'the greatest number are to enjoy the greatest possible development of the capacities of personality', considered such redistribution to be 'a matter for constant adjustment and readjustment, as social thought about justice grows and as the interpretation of the principles of liberty and equality broadens with its growth'.[29] And he thought that the adjustments now required 'may well begin, and may even sometimes remain, at the level of voluntary agreement between voluntary associations (those of the workers and those of the employers), an agreement based on voluntary consultation and issuing in voluntary co-operation.' When in this way something had been worked out that was 'so obviously best' as to deserve to be made a general rule, state action would be appropriate. 'In that case the State, which is not the enemy of Society, but rather stands to it in something of the relation in which a

[29] Ernest Barker: *Principles of Social & Political Theory*, Oxford, 1951, pp. 271–2.

solicitor may stand to a family, will register and endorse this best as a rule for general application and enforcement.'[30]

The notion that class differences could be adjusted 'as social thought about justice grows', and that this could be done by voluntary class co-operation aided by a family-solicitor state, is something of a retreat from Mill's appreciation of the class problem. It also makes Mill's utilitarian analysis appear hard-headed and realistic in comparison with the later idealists' reliance on goodwill.

In a similar vein, MacIver defined democratic states as those 'in which the general will is inclusive of the community as a whole or of at least the greater portion of the community, and is the conscious, direct, and active support of the form of government.'[31] He specifically distinguished democratic states from class-controlled states, and found that in modern civilizations classes shaded into one another and had 'no determinate solidarity of interest'.[32] He drew attention to the enormous range of interest groups and associations, making up a social universe where there is 'ceaseless motion and commotion, struggle and accord'.[33] And he saw the party system as the effective way of reducing 'the multitudinous differences of opinion to relatively simple alternatives'.[34] The task of the democratic state, a task which it did perform, however roughly, was to express and enforce the general will by representing men as citizens rather than as holders of particular interests.

> The danger is not that particular interests will not be focused and asserted but rather that the general interest may suffer domination through their urgency. Against this danger the chief bulwark is the state, because its organization presupposes and in some degree realizes the activity of the general will. Besides, we must assume that through the rough method of political representation the 'pluses and minuses' of particularist and opposing aims will, as Rousseau said, in a measure cancel out.

[30] Ibid., pp. 275–6.

[31] R. M. MacIver: *The Modern State*, Oxford, 1926, p. 342.

[32] Ibid., p. 403.

[33] MacIver: *The Web of Government*, New York, 1947, p. 435; cf. *Modern State*, p. 461.

[34] *Web of Government*, p. 214.

. . . Men are not content to be represented simply as farmers or as engineers or as Anglicans or as lovers of music or any other art or recreation: they want also to be represented as citizens. Otherwise the unity of their individual lives is unexpressed, no less than the unity of society. This representation is achieved, no matter how roughly, through the development of the party system. We have seen that though parties are dominated by strong particular interests they are in idea and in principle the formulations of the broader attitudes of citizenship. Unless they were, the state would fall to pieces.[35]

Thus MacIver offered his vision of the essential function of 'the state' as a description of the function actually performed, though imperfectly, by liberal-democratic states through their party systems.

When we turn from the neo-idealist view to John Dewey's pragmatist view of liberal democracies, we find it less indulgent about their actual operation. Yet he held out as a possibility and a hope what the idealist pluralists treated as an achievement. He had few illusions about the actual democratic system, or about the democratic quality of a society dominated by motives of individual and corporate gain. The root difficulty lay not in any defects in the machinery of government but in the fact that the democratic public was 'still largely inchoate and unorganized', and unable to see what forces of economic and technological organization it was up against.[36] There was no use tinkering with the political machinery: the prior problem was 'that of discovering the means by which a scattered, mobile and manifold public may so recognize itself as to define and express its interests'.[37] The public's present incompetence to do this was traced to its failure to understand the technological and scientific forces which had made it so helpless. The remedy was to be sought in more, and more widespread, social knowledge: 'democracy is a name for a life of free and enriching communion. It had its seer in Walt Whitman. It will have its consummation when free social enquiry is indissolubly wedded to the art of full and moving communication.'[38]

[35] *Modern State*, pp. 465–6.
[36] John Dewey: *The Public and Its Problems* (1927), Denver, 1954, p. 109.
[37] Ibid., p. 146.
[38] Ibid., p. 184.

What was needed was not just more education—a remedy to which many earlier liberals had had recourse—but an improvement in the social sciences by applying the experimental method and 'the method of co-operative intelligence'.[39] 'The essential need . . . is the improvement of the methods and conditions of debate, discussion and persuasion. That is *the* problem of the public. . . . this improvement depends essentially upon freeing and perfecting the processes of inquiry and of dissemination of their conclusions.[40]

Also needed was a large measure of social control of economic forces. Writing under the impact of the great depression, Dewey argued for 'a planned co-ordination of industrial development', preferably by voluntary agreement, perhaps by way of a 'co-ordinating and directive council in which captains of industry and finance would meet with representatives of labor and public officials to plan the regulation of industrial activity . . .'; in any case, 'the introduction of social responsibility into our business system to such an extent that the doom of an exclusively pecuniary-profit industry would follow.'[41] A few years later, denouncing 'control by the few of access to means of productive labor on the part of many', and noting 'the existence of class conflicts, amounting at times to veiled civil war', he argued that liberalism should go beyond the provision of social services 'and socialize the forces of production, now at hand, so that the liberty of individuals will be supported by the very structure of economic organization'.[42] But 'the forces of production' which were to be socialized were science and technology, which were now perverted from their proper end. This could not be done either by patchwork or by socialist revolution, but only by 'the method of cooperative intelligence'.[43] Although he referred more than once to the desirability of 'a socialized economy',[44] it is not at all clear what he had in mind. He was not interested in any analysis of

[39] *Liberalism and Social Action* (1935), New York, 1963, p. 81; cf. *Public and Its Problems*, p. 202.

[40] *Public and Its Problems*, p. 208.

[41] *Individualism Old and New* (1929), New York, 1962, pp. 117–18.

[42] *Liberalism and Social Action*, pp. 38, 80, 88.

[43] Ibid., p. 81

[44] Ibid., pp. 90, 91.

capitalism. He was entirely taken up with the prospects of a democratic liberalism. Acknowledging 'that our institutions, democratic in form, tend to favor in substance a privileged plutocracy', he went on to say:

Nevertheless it is sheer defeatism to assume in advance of actual trial that democratic political institutions are incapable either of further development or of constructive social application. Even as they now exist, the forms of representative government are potentially capable of expressing the public will when that assumes anything like unification.[45]

What above all was needed was for liberals to apply to 'social relations and social direction' the method of 'experimental and cooperative intelligence' that had already accomplished so much 'in subduing to potential human use the energies of physical nature'.[46]

Dewey, then, while far from relying on the existing democratic political machinery to bring about the desired transformation of society, appealed from democratic machinery to democratic humanism. Democracy 'is a way of life': it 'cannot now depend upon or be expressed in political institutions alone'.[47] The humanistic view which he saw as the essential of democracy must be infused into 'every phase of our culture—science, art, education, morals and religion, as well as politics and economics'.[48] This was to be done primarily through the spread of a scientific outlook: 'the future of democracy is allied with the spread of the scientific attitude.' And it must all be done by 'plural, partial and experimental methods'.[49]

The distance between Dewey's pragmatism, with its strong early twentieth-century influence in the United States, and the pluralist idealism which was so prevalent in English liberal-democratic thinking in the same period, is not great. Both saw a need for 'plural, partial and experimental methods'. The English theorists were more inclined to revert to the values of

[45] Ibid., pp. 85–6.

[46] Ibid., p. 92.

[47] *Freedom and Culture*, New York, 1939, pp. 130, 125.

[48] Ibid., p. 125.

[49] Ibid., pp. 148, 176.

ancient Athens, the Americans to the taming of technology; but both were firm believers in the efficacy of pluralism.

It is perhaps not unfair to say that all of them had unconsciously accepted the image of the democratic political process as a market, a free market in which everything would work out to the best advantage of everybody (or to the least disadvantage of anybody). They did not make the market analogy explicitly, because it was too crass, too materialistic: they still held to the democratic ideal of individual self-development, whereas the market analogy implied narrow seeking of immediate self-interest. They did not wish to impute to the citizen the narrow rationality of market man. But they could and did impute a citizen rationality capable of overcoming the imperfections of the actual democratic system. They were encouraged to do this because the actual system had survived: MacIver, for instance, could cite the fact of its survival as evidence that citizens had, in addition to their particular will, a rational general will as citizens, and that the system did allow that will to be expressed.[50] What the twentieth-century developmental theorists did not see, as we have noticed, was the extent to which the system had survived by reducing the responsiveness of governments to electorates. It was the developmental theorists' failure to see this that enabled them to postulate an overriding citizen rationality and build it into their descriptive model. And it was their putting this in their descriptive model that left them wide open to the shattering attack of the mid-twentieth-century empirical political scientists. In the end, it was the failure of the developmental theorists to see the difference between the actual democratic system which was very much like a market (although far from a fully competitive market), and their idealistic developmental hopes, that led to the failure of Model 2B and its supersession by Model 3, which was an entirely tough, and seemingly realistic, market model.

[50] As quoted above, at n. 35.

IV

Model 3: Equilibrium Democracy

THE ENTREPRENEURIAL MARKET ANALOGY

Model 3, the model which came to prevail in the Western world in the middle decades of the twentieth century, was offered as a replacement for the failed Model 2. It is, to an extent not always realized, a reversion to and elaboration of Model 1. That is the measure at once of its congruence with market society and bourgeois man, and of its increasingly apparent inadequacy.

I have called Model 3 the equilibrium model. It may equally well be called, as it sometimes is, the pluralist élitist model. Perhaps the only adequately descriptive name would be one which combined all three terms, 'the pluralist élitist equilibrium model', for these three characteristics are equally central to it. It is pluralist in that it starts from the assumption that the society which a modern democratic political system must fit is a plural society, that is, a society consisting of individuals each of whom is pulled in many directions by his many interests, now in company with one group of his fellows, now with another. It is élitist in that it assigns the main role in the political process to self-chosen groups of leaders. It is an equilibrium model in that it presents the democratic process as a system which maintains an equilibrium between the demand and supply of political goods.

Model 3 was first systematically, though briefly, formulated in 1942, by Joseph Schumpeter, in a few chapters of his influential book *Capitalism, Socialism, and Democracy*. Since then it has been built up and made apparently solid by the work of many political scientists who have amplified and supported it by a substantial amount of empirical investigation of how voters in Western democracies actually behave and how existing Western political systems actually respond to their behaviour.[1]

The main stipulations of this model are, first, that democracy is simply a mechanism for choosing and authorizing governments, not a kind of society nor a set of moral ends; and second, that the mechanism consists of a competition between two or more self-chosen sets of politicians (élites), arrayed in political parties, for the votes which will entitle them to rule until the next election. The voters' role is not to decide political issues and then choose representatives who will carry out those decisions: it is rather to choose the men who will do the deciding. Thus Schumpeter: 'the role of the people is to produce a government . . . the democratic method is that institutional arrangement for arriving at political decisions in which individuals acquire the power to decide by means of a competitive struggle for the people's vote.'[2] The individuals who so compete are, of course, the politicians. The citizens' role is simply to choose between sets of politicians periodically at election time. The citizens' ability thus to replace one government by another protects them from tyranny. And, to the extent that there is any difference in the platforms of the parties, or in the general lines of policy to be expected of each party as a government (on the basis of its record), the voters in choosing between parties register their desire for one batch of political goods rather than another. The purveyors of the batch which gets the most votes become the authorized rulers until the next election: they cannot tyrannize because there will be a next election.

Model 3 deliberately empties out the moral content which Model 2 had put into the idea of democracy. There is no nonsense about democracy as a vehicle for the improvement of

[1] Leading works are: Bernard R. Berelson, Paul F. Lazarsfeld, and William N. McPhee: *Voting*, Chicago, 1954; Robert A. Dahl: *A Preface to Democratic Theory*, Chicago, 1956; Dahl: *Who Governs?*, New Haven, 1961; Dahl: *Modern Political Analysis*, Englewood Cliffs, N.J., 1963; Gabriel A. Almond and Sidney Verba: *The Civic Culture*, Princeton, 1963.

[2] Joseph Schumpeter: *Capitalism, Socialism, and Democracy*, 2nd edn., New York and London, 1947, p. 269.

mankind. Participation is not a value in itself, nor even an instrumental value for the achievement of a higher, more socially conscious set of human beings. The purpose of democracy is to register the desires of people as they are, not to contribute to what they might be or might wish to be. Democracy is simply a market mechanism: the voters are the consumers; the politicians are the entrepreneurs. It is not surprising that the man who first proposed this model was an economist who had worked all his professional life with market models. Nor is it surprising that the political theorists (and then the publicists and the public) took up this model as a realistic one, for they also have lived and worked in a society permeated by market behaviour. Not only did the market model seem to correspond to, and hence to explain, the actual political behaviour of the main component parts of the political system—the voters and the parties; it also seemed to justify that behaviour, and hence the whole system.

For in the mid-twentieth century, when it still did not seem too naïve to talk about consumers' sovereignty in the economic market, it was easy to see a parallel in the political market: the political consumers were sovereign because they had a choice between the purveyors of packages of political goods. It was easy for the political theorists to make the same assumptions as the economic theorists. In the economic model, entrepreneurs and consumers were assumed to be rational maximizers of their own good, and to be operating in conditions of free competition in which all energies and resources were brought to the market, with the result that the market produced the optimum distribution of labour and capital and consumer goods. So in the political model, politicians and voters were assumed to be rational maximizers, and to be operating in conditions of free political competition, with the result that the market-like political system produced the optimum distribution of political energies and political goods. The democratic political market produced an optimum equilibrium of inputs and outputs—of the energies and resources people would put into it and the rewards they would get out of it. I have pointed out elsewhere[3] that by the time the political scientists had

[3] *Democratic Theory: Essays in Retrieval*, Oxford, 1973, Essay X.

taken over this economic model it was already being discarded or much modified by economists in favour of an oligopolistic power-bloc model of the economy. But the notion of consumers' sovereignty is still accepted in the pluralist political model, and serves as an implicit justification of it.

This model makes another market assumption. Not only does it assume that political man, like economic man, is essentially a consumer and an appropriator: it assumes also that the things different people want out of the government—the demands for political goods—are so diverse and shifting that the only way of making them effective, the only way of getting the government's decisions to meet them, the only way of eliciting the required supply of political goods and getting it distributed in proportion to the myriad demands, is an entrepreneurial system like that which operates in the standard model of the competitive market economy. Given that the political demands are so diverse that no natural or spontaneous grouping of them could be expected to produce a clear majority position, and given that in a democracy the government should express the will of the majority, it follows that a device is needed which will produce a majority will out of those diverse demands, or will produce the set of decisions most agreeable to, or least disagreeable to, the whole lot of diverse individual demands. A system of entrepreneurial political parties offering differently proportioned packages of political goods, of which the voters by majority vote choose one, is offered as the best, or the only, device for doing this: it produces a stable government which equilibrates demand and supply.

This pluralism of Model 3 evidently has something in common with the pluralism we have seen in Model 2B. But there is a considerable qualitative difference. The pluralism of Model 3 leaves out the ethical component that was so prominent in Model 2B. It treats citizens as simply political consumers, and political society as simply a market-like relation between them and the suppliers of political commodities.

From this summary account of Model 3 and the assumptions on which it is based, we can see that it offers itself as a statement of what the prevailing system actually is and as an explanation, in terms of market principles, of why it works as well as it does. We have noticed also that the explanation easily merges into justification. Before we look more closely at the adequacy of Model 3, as description, explanation, and justification, we should notice that there are differences of emphasis, if not of substance, between some of its leading exponents.

The differences are not so much in the descriptions they give as in the extent of the claims made for the system. They all see the citizens as political consumers, with very diverse wants and demands. They all see competition between politicians for the citizens' votes as the motor of the system. They all find that this mechanism does produce a stable equilibrium. They differ somewhat in their views of the extent to which it also provides some measure of political consumers' sovereignty. Schumpeter gives the system a rather low rating on this. He finds that the voters have most of their choices made for them,[4] and that the pressures they can bring to bear on the government between election times are not very effective.

Other analysts are more optimistic about the effectiveness of consumers' preferences. Dahl finds 'somewhat defective' in Schumpeter's 'otherwise excellent analysis' the view 'that elections and interelection activity are of trivial importance in determining policy'. But the most Dahl claims for these activities is that 'they are crucial processes for insuring that political leaders will be somewhat responsive to the preferences of some ordinary citizens';[5] or that 'With all its defects, [the American political system] does nonetheless provide a high probability that any active and legitimate group will make itself heard effectively at some stage in the process of decision . . . it appears to be a relatively efficient system for reinforcing agreement, encouraging moderation, and maintaining social peace in a restless and immoderate people operating a gigantic, powerful, diversified, and incredibly complex society.'[6] In a later work Dahl rates the responsiveness of the system a little higher: 'most citizens . . . possess a moderate degree of indirect influence, for elected officials keep the real or imagined preferences

[4] See below, at nn. 23 and 24.

[5] *Preface to Democratic Theory*, p. 131.

[6] Ibid., pp. 150–1.

of constituents constantly in mind in deciding what policies to adopt or reject.'[7]

Still higher claims are sometimes made. For instance, the influential study *Voting*, by Berelson, Lazarsfeld, and McPhee, after demonstrating that in the American political system the citizens are not at all like the rational citizens of Model 2, and pointing out that nevertheless the system does work (that is, has not disintegrated into either dictatorship or civil war), and 'often works with distinction',[8] concluded that it must have hidden merit. Something like the invisible hand celebrated by Adam Smith must be at work.

> If the democratic system depended solely on the qualifications of the individual voter, then it seems remarkable that democracy has survived through the centuries. After examining the detailed data on how individuals misperceive political reality, or respond to irrelevant social influences, one wonders how a democracy ever solves its political problems. But when one considers the data in a broader perspective—how huge sections of the society adapt to political conditions affecting them or how the political system adjusts itself to changing conditions over long periods of time—he cannot fail to be impressed with the total results. Where the rational citizen seems to abdicate, nevertheless angels seem to tread.[9]

This echo of Adam Smith is not surprising, for Berelson *et al.* do tend to attribute the success of Model 3 to its market-like nature: nothing less than the magic of the market can explain the success of the system, and nothing more is needed to justify it.

THE ADEQUACY OF MODEL 3

We have noticed that Model 3 presents itself as description, as explanation, and sometimes as justification, of the actual political system in Western democracies. In asking now how adequate the model is on each count we must acknowledge that there is some difficulty in treating the three counts separately, since they often merge into each other. Things may be left out of the descriptions because an explanatory framework

[7] *Who Governs?*, p. 164.

[8] Berelson, Lazarsfeld, and McPhee: *Voting*, p. 312.

[9] Ibid., p. 311.

already adopted treats them as of little or no importance. Or empirical descriptive findings about, for instance, citizens' apathy or voters' misinformation, may require the theorists to cast about for a principle of explanation to account for the fact that the system works at all. And principles of explanation, as we have seen, easily shade into justifications. One may still usefully separate the descriptive from the justificatory aspect, without hoping to treat the explanatory aspect entirely separately.

(i) *Descriptive adequacy*

As description of the actual system now prevailing in Western liberal-democratic nations, Model 3 must be adjudged substantially accurate. It is clearly a much more realistic statement than any provided by Model 2. It has been built up by careful and extensive empirical investigations by highly competent scholars. There is no reason to doubt their findings, which depart so drastically from Model 2. They may have left some things out of account, for instance the ability of the élites to decide what issues may be put to the voters at all and what are non-issues,[10] but such omissions may be thought to affect the model's explanatory or justificatory adequacy more than its descriptive adequacy.

Some adjustment may be needed to make their findings, which are pre-eminently based on researches into the system in the United States, applicable to Western Europe: the current strength of the Communist Party in France and Italy, for instance, suggests that in those countries party divisions are more polarized along class lines than the American pluralistic model allows for. But that can probably be accommodated without much difficulty. The substantial accuracy of Model 3 as description may be attributed to the substantial accuracy of its assumptions about current Western man and society: as long as we have market man and market society, they can be expected to operate as described in Model 3.

[10] As argued by Peter Bachrach and Morton S. Baratz: 'Two Faces of Power', *American Political Science Review*, LVI, 4 (December 1962); reprinted in Charles A. McCoy and John Playford (eds.): *Apolitical Politics, a Critique of Behavioralism*, New York, 1967.

(ii) *Explanatory adequacy*

Explanatory principles, intended to show why the system works at all or works as well as it does, grow out of (and grow into) the descriptive findings. But they also merge so generally into justifications of the system that it will be convenient to consider explanatory and justificatory adequacy together. Indeed, most of the recent writing criticizing Model 3 seems to have begun from dissatisfaction with its justificatory claims and gone on to challenge its explanatory or even its descriptive adequacy. I shall not attempt to summarize all the critical analyses of Model 3 that have been made in the last decade or so by political scientists of what may be called a radical liberal-democratic persuasion,[11] but simply cite their work as evidence of increasing dissatisfaction with the model among the political science community. I shall then go on to inquire, in the light of the analysis already made of the failure of Models 1 and 2, why Model 3 has begun to appear so unsatisfactory.

(iii) *Justificatory adequacy*

It may be well to begin by considering the claim generally made or implied by exponents cf Model 3 that their model is not justificatory at all, but only descriptive and explanatory. This claim really cannot be accepted, although Schumpeter, who scarcely bothered to make such a claim, might be justified in making it. But the later and more substantial exponents of Model 3 all imply, or even state, a justification at one or both of two levels. They are saying, at the least, that the system is, with all its admitted imperfections, the only one that can do the job, or the one that can do it best. They are the realists. That is what people are like, so this is the best they are capable of. Generally, even more is claimed—that the system produces optimum equilibrium and some measure of citizen consumers' sovereignty. These are taken to be self-evidently good, so the

[11] e.g. Peter Bachrach: *The Theory of Democratic Elitism, a Critique*, Boston and Toronto, 1967; McCoy and Playford, op. cit.; William Connolly (ed.): *The Bias of Pluralism*, New York, 1969; Henry Kariel (ed.): *Frontiers of Democratic Theory*, New York, 1970; Carole Pateman: *Participation and Democratic Theory*, Cambridge, 1970.

system which provides them is taken to be justified by the very demonstration that it does provide them. Both of the realists' claims are thus, at least implicitly, justificatory. How adequate are they?

The first claim amounts to saying that Model 3 is best because anything loftier is unworkable. The advocates of Model 3 contrast it with what they usually call the 'classical' model of democracy, which generally turns out to be a confused mixture of a pre-industrial model (Rousseau's or Jefferson's), and our Models 1 and 2. It would take too long a digression to try to sort out those confusions,[12] especially as different proponents of Model 3 set up their 'classical' straw men rather differently. Schumpeter, for instance, makes his main target the over-rationalistic assumptions he finds in Rousseau and in Bentham's Model 1: average men, he holds, are not capable of forming the rational judgements he thinks required by those models; therefore those models are hopeless.[13] Others have been more concerned to deflate the moral pretensions of Model 2, while accepting the Model 1 view of man as essentially a rational maximizing calculator: it is *because* men are on the whole such maximizing calculators that most of them may well decide not to spend much time or energy in political participation, thus invalidating Model 2.[14]

[12] The extent of the confusion has been pointedly remarked by Carole Pateman: 'the notion of a "classical theory of democracy" is a myth' (*Participation and Democratic Theory*, p. 17).

[13] A similar although less extravagant position is taken by Berelson (Berelson, Lazarsfeld, and McPhee: *Voting*, p. 322).

[14] Cf. Robert Dahl's argument (*After the Revolution? Authority in a Good Society*, New Haven and London, 1970, pp. 40–56) that 'a reasonable man will' and 'in actual practice everyone does' apply, to any system of authority, the 'Criterion of Economy', which is to balance the cost of political participation against the expected benefit, the cost being the forgone uses of his time and energy. This notion of participation as nothing but a 'cost' (which it is, if everyone is seen as merely a maximizing consumer) overlooks the possible value of participation in enhancing the participant's understanding of his own position and in giving him a greater sense of purpose and greater awareness of community. Cf. Bachrach: 'Interest, Participation, and Democratic Theory', in J. R. Pennock and J. W. Chapman (eds.): *Participation in Politics* (Nomos XVI), New York, 1973, pp. 49–52.

Both these views as to why Model 3 is more realistic, more workable, and so 'better', than any previous model, rest ultimately on an unverifiable assumption that the political capabilities of the average person in a modern market society are a fixed datum, or at least are unlikely to change in our time.

One might argue, against the validity of that assumption, that it depends on a model of man which came to prevail only with the emergence or predominance of the capitalist market society.[15] But even if it is granted that that model of man is so time-bound and culture-bound, we do not know whether or when it may be superseded. So, although the assumption cannot be verified, neither can it be absolutely falsified. Hence the justificatory adequacy of the first claim must be left undecided: we can only return the Scottish verdict 'Not Proven'.

What of the second claim: that, on the analogy of the market in the economic system, the competitive élite party system brings about an optimum equilibrium of the supply and demand for political goods, and provides some measure of citizen consumer sovereignty? *Prima facie*, optimum equilibrium and citizen consumer sovereignty are good in themselves. To most people who live in advanced and relatively stable societies, 'equilibrium' sounds better than 'disequilibrium'; and 'optimum' is by definition best; so what could be better than 'optimum equilibrium'? And 'citizen consumer sovereignty' is a phrase loaded with good words. So if Model 3 does provide these, surely we might conclude that it is a pretty good kind of democracy. But this does not follow. All that follows is that it is a pretty good kind of a market. But a market is not necessarily democratic.

I want now to show that the Model 3 political market system is not nearly as democratic as it is made out to be: that the equilibrium it produces is an equilibrium in inequality; that the consumer sovereignty it claims to provide is to a large extent an illusion; and that, to the extent that the consumer sovereignty is real, it is a contradiction of the central democratic tenet of equality of individual entitlement to the use and enjoyment of one's capacities. The claims for optimum equi-

[15] Cf. Karl Polanyi: *The Great Transformation*, New York, 1944, and my *Democratic Theory*, Essay I.

librium and consumer sovereignty are virtually the same claim—two sides of the same coin—and so may be treated together as a single claim.

The claim fails on two counts. First, in so far as the political market system, on the analogy of the economic market, is competitive enough to produce the optimum supply and distribution of political goods, optimum in relation to the demands, what it does is to register and respond to what economists call the *effective* demand, that is, the demands that have purchasing power to back them. In the economic market this means simply money, no matter whether the money has been acquired by an output of its possessors' energy or in some other way. In the political market the purchasing power is to a large extent, but not entirely, money—the money needed to support a party or a candidate in an election campaign, to organize a pressure group, or to buy space or time in the mass media (or to own some of the mass media). But political purchasing power includes also direct expenditure of energy in campaigning, organizing, and participating in other ways in the political process.

In so far as the political purchasing power is money, we can scarcely say that the equilibrating process is democratic in any society, like ours, in which there is substantial inequality of wealth and of chances of acquiring wealth. We may still call it consumer sovereignty if we wish. But the sovereignty of an aggregate of such unequal consumers is not evidently democratic.

In so far as the political purchasing power is direct expenditure of energy the case seems better. What could be fairer than a return proportional to the input of political energy? Citizens who are apathetic should surely not expect as much return as those who are more active. This would be a fair principle, consistent with democratic equality, if the apathy were an independent datum, that is, if the apathy were in each case the outcome of a maximizing decision by the individual, balancing the most profitable uses of his time and energy as between political participation and other things, *and if* every individual could expect that each hour he gave to politics would have the same value, the same purchasing power in the political market, as any other person's. But this is just what it cannot have. Those whose education and occupation make it more difficult for them than for the others to acquire and marshal and weigh the information needed for effective participation are clearly at a disadvantage: an hour of their time devoted to political participation will not have as much effect as an hour of one of the others. They know this, hence they are apathetic. Social inequality thus creates political apathy. Apathy is not an independent datum.

Over and above this, the political system of Model 3 contributes directly to apathy. As we saw in the preceding chapter, the functions which a party system in an unequal society with mass franchise must perform require a blurring of issues and a diminution of the responsibility of governments to electorates, both of which reduce the incentive of the voters to exert themselves in making a choice. A frequent reason for non-voting is the feeling that there is no real choice.

Proponents of Model 3 have made much of the phenomenon of voter apathy, though they have not usually traced it to the causes I have just mentioned. They do, however, often point out that successful operation of Model 3 *requires* something like the present levels of apathy: greater participation would endanger the stability of the system.[16] The accuracy of this general proposition is never demonstrated, but the fact that it is asserted at all is revealing: in the realism of Model 3, some good is to be found even in something as unpromising as widespread apathy. We may prefer to think that a political system which requires and encourages apathy is not doing a very brisk job of optimizing, especially in view of the class differential in apathy.[17]

To sum up, then, on the first count, we find that in so far as

[16] e.g. Berelson *et al.*: *Voting*, ch. 14; W. H. Morris-Jones: 'In Defence of Apathy', *Political Studies* II (1954), pp. 25–37; Seymour Martin Lipset: *Political Man*, New York, 1960, pp. 14–16; Lester W. Milbrath: *Political Participation*, Chicago, 1965, ch. 6.

[17] That there is a class differential in political participation is the unanimous conclusion of voting studies. For a thorough exploration of this and other dimensions of apathy, see Sidney Verba and Norman H. Nie: *Participation in America, Political Democracy and Social Equality*, New York, 1972.

the political market system *is* competitive enough to do the job of equilibrating the supply of and demand for political goods—in so far, that is, as it *does* actually respond to consumer demands—it measures and responds to demands which are very unequally effective. Some demands are more effective than others because, where the demand is expressed in human energy input, one person's energy input cannot get the same return per unit as another person's. And the class of political demands that have the most money to back them is largely the same as the class of those that have the larger pay-off per unit of human energy input. In both cases it is the demands of the higher socio-economic classes which are the most effective. So the lower classes are apathetic. In short, the equilibrium and the consumer sovereignty, in so far as Model 3 does provide them, are far from democratic.[18]

The second count on which the claim to provide a democratic consumer sovereignty fails is simply that Model 3 does not provide a significant amount of consumer sovereignty. The Model 3 political market is far from fully competitive. For it is, to use an economists' term, oligopolistic. That is, there are only a few sellers, a few suppliers of political goods, in other words only a few political parties: in the most favoured variant of Model 3 there are only two effective parties, with a possibility of one or two more. Where there are so few sellers, they need not and do not respond to the buyers' demands as they must do in a fully competitive system. They can set prices and set the range of goods that will be offered. More than that, they can, to a considerable extent, create the demand. In an oligopolistic market, the demand is not autonomous, not an independent datum.

This effect of oligopoly, which is a commonplace of economic theory, has been surprisingly little noticed by the political theorists of Model 3. Even Schumpeter, who of all the formulators of Model 3 has economic parallels most in mind, and who makes quite a point of the way that oligopoly and imperfect competition require a substantial revision of the classical

[18] Dahl, who has explored the implications of Model 3 more fully than most of its exponents, particularly in his *After the Revolution* (1970), is there explicit about the distorting effect of class inequality and sees its reduction as a prerequisite of genuine democracy.

and neo-classical economic theory of equilibrium, does not see its importance in his political model. He mentions the parallel between economic and political imperfect competition,[19] but it is imperfect competition of all degrees, not the highly imperfect form which is oligopoly, that he has in mind: instead of dealing with the crucial fact of party oligopoly, he describes 'party and machine politicians' as 'an attempt to regulate political competition exactly similar to the corresponding practices of a trade association'.[20]

Why should the ability of oligopolistic parties to create the demands for political goods have been so overlooked? It is, I think, because the theorists had already postulated that, quite apart from the extent of party competition, the voters' demands are not and cannot be the independent ultimate data of the political system.[21] This follows from their prior postulate that the democratic party system is essentially a competition between élites. The élites being the driving force, they formulate the issues. So Schumpeter: 'what we are confronted with in the analysis of political processes is largely not a genuine but a manufactured will', manufactured in ways 'exactly analogous to the ways of commercial advertising';[22] the people 'neither raise nor decide issues but . . . the issues that shape their fate are normally raised and decided for them';[23] the wishes of the electorate 'are not the ultimate data', the electorate's choice 'does not flow from its initiative but is being shaped, and the shaping of it is an essential part of the democratic process'.[24]

Thus Model 3 asserts that, irrespective of the degree of oligopoly in the competition of parties, but due simply to the fact that the initiative is always in the élites, the basic, irreducible unit in the democratic process is *not* the individual with an independent, autonomous set of demands, or as economists would say, an autonomous demand schedule. Model 3 asserts instead that the demand schedule for political goods is itself

[19] *Capitalism, Socialism, and Democracy*, p. 271.
[20] Ibid., p. 283.
[21] Ibid., p. 254; cf. next three notes.
[22] Ibid., p. 263.
[23] Ibid., p. 264.
[24] Ibid., p. 282.

largely dictated by the suppliers. The assertion is accurate enough. But curiously, this fact is held not to invalidate Model 3's claim to be democratic, but to reinforce it. The argument is, that since individual demand schedules are *not* the independent basic data of the system, therefore the democratic process cannot hope to live up to the democratic expectations or ideal of Models 1 and 2, cannot hope to perform the functions attributed to it by Models 1 and 2 or by any variant of the 'classical' model, all of which depended on autonomous individuals: therefore, Model 3 is better than Models 1 and 2.

Now this perception, by the builders of Model 3, of the actual relations that prevail in our society, does reinforce the claim of Model 3 to be realistic—realistic, that is, for a society which is thought incapable of going beyond the oligopolistic economic market, the inequality of classes, and people's vision of themselves as essentially consumers. But it puts some strain on the claim that Model 3 is democratic. Since Model 3 permits, or even requires, that the élite suppliers of political goods have a large part in creating the demands (as they do, and must do, in an oligopolistic market), the bottom is knocked out of the optimum equilibrium and consumers' sovereignty case for Model 3. Little remains of the case for Model 3 except the sheer protection-against-tyranny function.

Now certainly no liberal, indeed no maximizing individual, will belittle the importance of protection against tyranny. If Model 3 were the only alternative to a dictatorship of an irremovable set of rulers, the case for Model 3, with all its inequality, oligopoly, and apathy, would still be compelling. But that Model 3 is the only alternative has never been demonstrated; indeed it is hardly ever explicitly argued. What is needed now is further inquiry into the possibility of a non-dictatorial system which would not have all the shortcomings of Model 3.

THE FALTERING OF MODEL 3

Model 3 will remain the most accurate descriptive model, and will continue to be accepted as an adequate justificatory model, as long as we in Western societies continue to prefer affluence to community (and to believe that the market society can provide affluence indefinitely), and as long as we continue to accept the cold-war view that the only alternative to Model 3 is a wholly non-liberal totalitarian state. Putting this in a slightly different way, we might say that a system of competing élites with a low level of citizen participation is *required* in an unequal society, most of whose members think of themselves as maximizing consumers.

This requirement took on a new urgency with the catastrophic economic depression of the early 1930s in all the Western nations. The need for the state to intervene in the economy along Keynesian lines, in order to sustain the capitalist order, meant an increased need to remove political decisions from any democratic responsiveness: only the experts, whose reasoning was assumed to be beyond the comprehension of the voters, could save the system. The experts' advice was followed, and it did save the system for the next three or four decades. Model 3 was, therefore, from its very beginnings in the 1940s, understandably aligned against democratic participation. But with increasing disillusionment with the results of this state-regulated capitalism in the 1960s and 70s, the adequacy of Model 3 is increasingly questioned.

The fact that doubts are increasingly being raised about the adequacy of this system cannot, unfortunately, be taken as evidence that we have moved far enough away from inequality, and from the consciousness of ourselves as essentially consumers, to make a new political model possible. The most we can do is to look at the problems of moving to a new model, and examine possible solutions.

V

Model 4: Participatory Democracy

THE RISE OF THE IDEA

To call participatory democracy a model at all, let alone a model of liberal democracy, is perhaps to yield too much to a liking for symmetry. Participatory democracy is certainly not a model as solid or specific as those we have been examining. It began as a slogan of the New Left student movements of the 1960s. It spread into the working class in the 1960s and '70s, no doubt as an offshoot of the growing job-dissatisfaction among both blue- and white-collar workers and the more widespread feeling of alienation, which then became such fashionable subjects for sociologists, management experts, government commissions of inquiry, and popular journalists. One manifestation of this new spirit was the rise of movements for workers' control in industry. In the same decades, the idea that there should be substantial citizen participation in *government* decision-making spread so widely that national governments began enrolling themselves, at least verbally, under the participatory banner, and some even initiated programmes embodying extensive citizen participation.[1] It appears that the hope of a more participatory society and system of government has come to stay.

We need not attempt to review the voluminous recent literature on participation in various spheres of society. Our concern

[1] e.g. the Community Action Programs inaugurated by the United States federal government in 1964, which called for 'maximum feasible participation of residents of the areas and members of the groups served'. For a critical account of this, see 'Citizen Participation in Emerging Social Institutions' by Howard I. Kalodner, in *Participation in Politics*, as cited in n. 3, below.

here is only with the prospects of a more participatory system of government for Western liberal-democratic nations. Can liberal-democratic government be made more participatory, and if so, how? This question has not yet had as much attention as it deserves. The debate among political theorists had to be at the beginning mainly concerned with the prior question: is more citizen participation desirable?[2] The exponents of Model 3, as we have seen, said no. That debate is not yet ended.[3]

For our purposes, however, that debate may be foreclosed. It is sufficient to say that in view of the unquestioned class differential in political participation in the present system, and assuming that that differential is both the effect and the continuing cause of the inability of those in the lower strata either to articulate their wants or to make their demands effective, then nothing as unparticipatory as the apathetic equilibrium of Model 3 measures up to the ethical requirements of democracy. This is not to say that a more participatory system would of itself remove all the inequities of our society. It is only to say that low participation and social inequity are so bound up with each other that a more equitable and humane society requires a more participatory political system.

The difficult question, whether either a change in the political system or a change in the society is a prerequisite of the other, will occupy us largely in the next section of this chapter. In the meantime I shall assume that something more participatory than our present system is desirable. The remaining question is whether it is possible.

IS MORE PARTICIPATION NOW POSSIBLE?

(i) *The problem of size*

It is not much use simply celebrating the democratic quality

[2] This has been the main concern of the radical liberal critics of Model 3 (as cited in ch. IV, p. 84, n. 11, and in n. 3, below.

[3] See *Participation in Politics* (Nomos XVI) (eds. J. R. Pennock and J. W. Chapman), New York, 1975. Most of the contributors to this volume, which is based on papers given at the 1971 annual meeting of the American Society for Political and Legal Philosophy, are in favour of more participation, but there is a spirited defence, by M. B. E. Smith, of the opposite position.

of life and of decision-making (that is, of government) that can be had in contemporary communes or New England town-meetings or that was had in ancient city-states. There may be a lot to learn about the quality of democracy by examining these face-to-face societies, but that will not show us how a participatory democracy could operate in a modern nation of twenty million or two hundred million people. It seems clear that, at the national level, there will have to be some kind of representative system, not completely direct democracy.

The idea that recent and expected advances in computer technology and telecommunications will make it possible to achieve direct democracy at the required million-fold level is attractive not only to technologists but also to social theorists and political philosophers.[4] But it does not pay enough attention to an inescapable requirement of any decision-making process: somebody must formulate the questions.

No doubt something could be done with two-way television to draw more people into more active political discussion. And no doubt it is technically feasible to put in every living-room—or, to cover the whole population, beside every bed—a computer console with Yes/No buttons, or buttons for Agree/Disagree/Don't Know, or for Strongly Approve/Mildly Approve/Don't Care/Mildly Disapprove/Strongly Disapprove, or for preferential multiple choices. But it seems inevitable that some government body would have to decide what questions would be asked: this could scarcely be left to private bodies.

There might indeed be a provision that some stated number of citizens have the right to propose questions which must then be put electronically to the whole electorate. But even with such a provision, most of the questions that would need to be asked in our present complex societies could scarcely be formulated by citizen groups specifically enough for the answers to give a government a clear directive. Nor can the ordinary citizen be expected to respond to the sort of questions that would be required to give a clear directive. The questions would have to be as intricate as, for instance, 'what per cent

[4] See Michael Rossman: *On Learning and Social Change*, New York, 1972, pp. 257–8; and Robert Paul Wolff: *In Defense of Anarchism*, New York, 1970, pp. 34–7.

unemployment rate would you accept in order to reduce the rate of inflation by x per cent?', or 'what increase in the rate of (a) income tax, (b) sales and excise taxes, (c) other taxes (specify which), would you accept in order to increase by blank per cent (fill in [punch in] the blank), the level of (1) old-age pensions, (2) health services, (3) other social services (specify which), (4) any other benefits (specify which)?' Thus even if there were provision for such a scheme of popular initiative, governments would still have to make a lot of the real decisions.

Moreover, unless there were, somewhere in the system, a body whose duty was to reconcile inconsistent demands presented by the buttons, the system would soon break down. If such a system were to be attempted in anything like our present society there would almost certainly be inconsistent demands. People—the same people—would, for instance, very likely demand a reduction of unemployment at the same time as they were demanding a reduction of inflation, or an increase in government expenditures along with a decrease in taxes. And of course different people—people with opposed interests, such as the presently privileged and the unprivileged—would also present incompatible demands. The computer could easily deal with the latter incompatibilities by ascertaining the majority position, but it could not sort out the former. To avoid the need for a body to adjust such incompatible demands to each other the questions would have to be framed in a way that would require of each voter a degree of sophistication impossible to expect.

Nor would the situation be any better in any foreseeable future society. It is true that the sort of questions just mentioned, which are about the distribution of economic costs and economic benefits among different sections of the population, may be expected to become less acute in the measure that material scarcity becomes less pressing. But even if they were to disappear as internal problems in the economically most advanced societies, they would reappear there as external problems: for instance, how much and what kind of aid should the advanced countries afford to the underdeveloped ones? Moreover, another range of questions would arise internally, having to do not with distribution but with production in the broadest sense, that is, with the uses to be made of the society's whole stock of energy and resources, and the encouragement or discouragement of further economic growth and population growth. And beyond that there would be such questions as the extent to which the society should promote or should keep its hands off the cultural and educational pursuits of the people.

Such questions, even in the most favourable circumstances imaginable, will require repeated reformulation. And questions of this sort do not readily lend themselves to formulation by popular initiative. Their formulation would have to be entrusted to a governmental body.

It might still be argued that even if it is impossible to leave the formulation of all policy questions to popular initiative, at least the very broadest sort of policy could be left to it. Granted that the many hundreds of policy decisions that are now made every year by governments and legislatures would still have to be made by them, it might be urged that those decisions should be required to conform to the results of referenda on the very broadest questions. But it is difficult to see how most of the broadest questions could be left to formulation by popular initiative. Popular initiative could certainly formulate clear questions on certain single issues, for instance, capital punishment or legalization of marijuana or of abortion on demand—issues on which the response required is simply yes or no. But for the reasons given above, popular initiative could not formulate adequate questions on the great interrelated issues of overall social and economic policy. That would have to be left to some organ of government. And unless that organ were either an elected body or responsible to an elected body, and thus at some remove responsible to the electorate, such a system of continual referenda would not really be democratic: worse, by giving the appearance of being democratic, the system would conceal the real location of power and would thus enable 'democratic' governments to be more autocratic than they are now. We cannot do without elected politicians. We must rely, though we need not rely exclusively, on indirect democracy. The problem is to make the elected politicians responsible. The electronic console beside every bed cannot do that. Electronic technology, then, cannot give us direct democracy.

So the problem of participatory democracy on a mass scale seems intractable. It *is* intractable if we simply try to draw mechanical blue-prints of the proposed political system without paying attention to the changes in society, and in people's consciousness of themselves, which a little thought will show must precede or accompany the attainment of anything like participatory democracy. I want to suggest now that the central problem is not how a participatory democracy would operate but how we could move towards it.

(ii) *A vicious circle and possible loopholes*

I begin with a general proposition: the main problem about participatory democracy is not how to run it but how to reach it. For it seems likely that if we can reach it, or reach any substantial instalment of it, our way along the road to reaching it will have made us capable of running it, or at least less incapable than we now are.

Having announced this proposition, I must immediately qualify it. The failures so far to reach really participatory democracy in countries where that has been a conscious goal, for instance Czechoslovakia in the years up to 1968 and many of the Third World countries, demand some reservations about such a proposition. For in both those cases, a good deal of the road had already been travelled: I mean the road away from capitalist class-division and bourgeois ideology towards, in the one case, a Marxist humanism and, in the other, a Rousseauan concept of a society embodying a general will, and in both cases towards a stronger sense of community than we have. And, of course, the whole of the road had there been travelled away from that mirror-image of the oligopolistic capitalist market system: I mean, the oligopolistic competition of political parties which prevails with us, which is not only not very participatory, but is recommended, by most current liberal-democratic theorists, as quintessentially non-participatory.

So there still are difficulties in reaching participatory democracy, even when much of the road has been travelled, i.e. when some of the obvious prerequisite changes in society and ideology *have* taken place. However, the roads they have travelled in such countries as I have just mentioned are significantly different from the road we would have to travel to come near to participatory democracy. For I assume that our road in the Western liberal democracies is not likely to be via communist revolution; nor, obviously, will it be via revolutions of national independence beset by all the problems of underdevelopment and low productivity that have faced the Third World countries.

It therefore seems worth inquiring what road it may be possible for any of the Western liberal democracies to travel, and whether, or to what extent, moving along that road could make us capable of operating a system substantially more participatory than our present one. This becomes the question: what roadblocks have to be removed, i.e. what changes in our present society and the now prevailing ideology are prerequisite or co-requisite conditions for reaching a participatory democracy?

If my earlier analysis is at all valid, the present nonparticipatory or scarcely participatory political system of Model 3 does fit an unequal society of conflicting consumers and appropriators: indeed, nothing but that system, with its competing political élites and voter apathy, seems competent to hold such a society together. If that is so, two prerequisites for the emergence of a Model 4 are fairly clearly indicated.

One is a change in people's consciousness (or unconsciousness), from seeing themselves and acting as essentially consumers to seeing themselves and acting as exerters and enjoyers of the exertion and development of their own capacities. This is requisite not only to the emergence but also to the operation of a participatory democracy. For the latter self-image brings with it a sense of community which the former does not. One can acquire and consume by oneself, for one's own satisfaction or to show one's superiority to others: this does not require or foster a sense of community; whereas the enjoyment and development of one's capacities is to be done for the most part in conjunction with others, in some relation of community. And it will not be doubted that the operation of a participatory democracy would require a stronger sense of community than now prevails.

The other prerequisite is a great reduction of the present social and economic inequality, since that inequality, as I have argued, requires a non-participatory party system to hold the society together. And as long as inequality is accepted, the non-participatory political system is likely also to be accepted by all those in all classes who prefer stability to the prospect of complete social breakdown.

Now if these two changes in society—the replacement of the image of man as consumer, and a great reduction of social and economic inequality—are prerequisites of participatory democracy, we seem to be caught in a vicious circle. For it is unlikely that either of these prerequisite changes could be effected without a great deal more democratic participation than there is now. The reduction of social and economic inequality is unlikely without strong democratic action. And it would seem, whether we follow Mill or Marx, that only through actual involvement in joint political action can people transcend their consciousness of themselves as consumers and appropriators. Hence the vicious circle: we cannot achieve more democratic participation without a prior change in social inequality and in consciousness, but we cannot achieve the changes in social inequality and consciousness without a prior increase in democratic participation.

Is there any way out? I think there may be, though in our affluent capitalist societies it is unlikely to follow the pattern proposed or expected in the nineteenth century either by Marx or by Mill. Marx expected the development of capitalism to lead to a sharpening of class consciousness, which would lead to various kinds of working-class political action, which would further increase the class consciousness of the working class and turn it into revolutionary consciousness and revolutionary organization. This would be followed by a revolutionary take-over of power by the working class, which power would be consolidated by a period of 'dictatorship of the proletariat', which would break down the social and economic inequality and replace man as maximizing consumer by man as exerter and developer of his human capacities. Whatever we may think of the probability of this sequence once it had started, it does require increasing class consciousness to start it, and there is little evidence of this in prosperous Western societies today, where it has generally declined since Marx's day.[5]

John Stuart Mill's way out does not seem very hopeful either. He counted on two things. First, the broadening of the franchise would lead to more widespread political participation which would in turn make people capable of still more political participation and would contribute to a change in consciousness. Secondly, the owner/worker relation would change with the spread of producers' co-ops: to the extent that they replaced the standard capitalist relation, both consciousness and inequality would be changed. But the broadening of the franchise did not have the result Mill hoped for, nor has the capitalist relation between owner and worker changed in the way required.

So neither Marx's nor Mill's way seems a way out of our vicious circle. But there is one insight common to both of them that we might well follow. Both assumed that changes in the two factors which abstractly seem to be prerequisites of each other—the amount of political participation on the one hand, and the prevailing inequality and the image of man as infinite consumer and appropriator on the other—would come stage by stage and reciprocally, an incomplete change in one leading to some change in the other, leading to more change in the first, and so on. Even Marx's scenario, including as it did revolutionary change at one point, called for this reciprocal incremental change both before and after the revolution. We also may surely assume, in looking at our vicious circle, that we needn't expect one of the changes to be complete before the other can begin.

So we may look for loopholes anywhere in the circle, that is, for changes already visible or in prospect either in the amount of democratic participation or in social inequality or consumer consciousness. If we find changes which are not only already perceptible but which are attributable to forces or circumstances which are likely to go on operating with cumulative

[5] There are some signs that class consciousness is re-emerging (see below, p. 106), but not that it is becoming a revolutionary consciousness.

effect, then we can have some hope of a break-through. And if the changes are of a sort that encourages reciprocal changes in the other factors, so much the better.

Are there any loopholes which come up to these specifications? Let us start from the assumption least favourable to our search, the assumption that most of us are, willy-nilly, maximizing calculators of our own benefit, making a cost/benefit analysis of everything, however vaguely we make it; and that most of us consciously or unconsciously see ourselves as essentially infinite consumers. From these assumptions the vicious circle appears to follow directly: most people will support, or not do much to change, a system which produces affluence, which continually increases the Gross National Product, and which also produces political apathy. This makes a pretty strong vicious circle. But there are now some visible loopholes. I shall draw attention to three of them.

(1) More and more people, in the capacity we have attributed to them all, namely as cost/benefit calculators, are reconsidering the cost/benefit ratio of our society's worship of expansion of the GNP. They still see the benefits of economic growth, but they are now beginning to see some costs they hadn't counted before. The most obvious of these are the costs of air, water, and earth pollution. These are costs largely in terms of the quality of life. Is it too much to suggest that this awareness of quality is a first step away from being satisfied with quantity, and so a first step away from seeing ourselves as infinite consumers, towards valuing our ability to exert our energies and capacities in a decent environment? Perhaps it is too much. But at any rate the growing consciousness of these costs weakens the unthinking acceptance of the GNP as the criterion of social good.

Other costs of economic growth, notably the extravagant depletion of natural resources and the likelihood of irreversible ecological damage, are also increasingly being noticed. Awareness of the costs of economic growth takes people beyond sheer consumer consciousness. It can be expected to set up some consciousness of a public interest that is not looked after either by the private interest of each consumer or by the competition of political élites.

(2) There is an increasing awareness of the costs of political apathy, and, closely related to this, a growing awareness, within the industrial working class, of the inadequacy of traditional and routine forms of industrial action. It is coming to be seen that citizens' and workers' non-participation, or low participation, or participation only in routine channels, allows the concentration of corporate power to dominate our neighbourhoods, our jobs, our security, and the quality of life at work and at home. Two examples of this new awareness may be given.

(a) The one that is most evident, at least in North American cities, which have hitherto been notoriously careless of human values, is the rise of neighbourhood and community movements and associations formed to exert pressure to preserve or enhance those values against the operations of what may be called the urban commercial-political complex. Such movements have sprung up, with substantial effect, against expressways, against property developers, against inner-city decay, for better schools and day-care centres in the inner city, and so on. It is true that they have generally begun as, and sometimes remained, single-issue affairs. And they do not usually seek to replace, but only to put new pressures on, the formal municipal political structure.[6] Most of them do not, therefore, by themselves constitute a significant breakaway from the system of competing élites. But they do attract to active political participation many, especially of the lower socio-economic strata, who had previously been most politically apathetic.

(b) Less noticeable, but probably in the long run more important, are the movements for democratic participation in decision-making at the workplace. These movements have not yet made decisive strides in any of the capitalist democracies, but pressure for some degrees of workers' control at the shop-floor level and even at the level of the firm is increasing, and

[6] Sometimes they do seek to revise the formal structure, as in the demands for community control of schools or police and for greater community participation in city planning and intelligence operations, as mentioned by John Ladd: 'The Ethics of Participation', in J. R. Pennock and J. W. Chapman, op. cit., pp. 99, 102.

examples of it actually in operation are promising.[7] The importance of this, whether the decisions are only about working conditions and planning the way the work is to be arranged at the shop-floor level, or whether it goes as far as participation in policy decisions at the level of the firm, is twofold.

In the first place, those who are involved in it are getting experience of participation in decision-making in that side of their lives—their lives at work—where their concern is greater, or at least more immediately and directly felt, than in any other. They can see at first hand just how far their participation is effective. The forces which make for the apathy of the ordinary person in the formal political process of a whole nation are absent. Unconcern about the outcome of apparently far-off political issues; distance from the results, if any, of participation; uncertainty about or disbelief in the effectiveness of their participation; lack of confidence in their own ability to participate—none of these apply to participation in decisions at the workplace. And an appetite for participation, based on the very experience of it, may well carry over from the workplace to wider political areas. Those who have proved their competence in the one kind of participation, and gained confidence there that they can be effective, will be less put off by the forces which have kept them politically apathetic, more able to reason at a greater political distance from results, and more able to see the importance of decisions at several removes from their most immediate concerns.

In the second place, those involved in workers' control are participating *as producers*, not as consumers or appropriators.

[7] An effective analysis of these is given by Carole Pateman: *Participation and Democratic Theory*, Cambridge, 1970, chs. 3 and 4. Other analysts, writing as political activists who want workers' control as a path to a fully socialist society, find the present achievement of the workers' control movements less encouraging, e.g. Gerry Hunnius, G. D. Garson, and John Case (eds.): *Workers' Control, a Reader on Labor and Social Change*, New York, 1973; and Ken Coates and Tony Topham (eds.): *Workers' Control, a book of readings and witnesses for workers' control*, London, 1970. The pressure for workers' control is likely to increase since it flows from the increasing, degradation of work which seems inherent in capitalist production: cf. Harry Braverman: *Labour and Monopoly Capital: the Degradation of Work in the Twentieth Century*, New York and London, 1974.

They are in it not to get a higher wage or a greater share of the product, but to make their productive work more meaningful to them. If workers' control were merely another move in the scramble for more pay to take home, or in the continuing effort to maintain real wages by getting increased money wages and fringe benefits, which is what much trade union activity is about, it would do nothing, just as established trade union practice does nothing, to move men away from their image of themselves as consumers and appropriators. But workers' control is not primarily about distribution of income: it is about the conditions of production, and as such it can be expected to have a considerable breakaway effect.

(3) There is a growing doubt about the ability of corporate capitalism, however much aided and managed by the liberal state, to meet consumer expectations in the old way, i.e. with the present degree of inequality. There is a real basis for this doubt: the basis is the existence of a contradiction within capitalism, the results of which cannot be indefinitely avoided.

Capitalism reproduces inequality and consumer consciousness, and must do so to go on operating. But its increasing ability to produce goods and leisure has as its obverse its increasing need to spread them more widely. If people can't buy the goods, no profit can be made by producing them. This dilemma can be staved off for quite a time by keeping up cold war and colonial wars: as long as the public will support these, then the public is, as consumers, buying by proxy all that can be profitably produced, and is wasting it satisfactorily. This has been going on for a long time now, but there is at least a prospect that it will not be indefinitely supported as normal. If it is not supported, then the system will either have to spread real goods more widely, which will reduce social inequality; or it will break down, and so be unable to continue to reproduce inequality and consumer consciousness.

This dilemma of capitalism is much more intense now than it was in the nineteenth century, when capitalism had the big safety-valves of continental and colonial expansion. The dilemma, in conjunction with the changing public awareness of the cost/benefit ratio of the system, puts capitalism in a rather different position from the one it enjoyed in Mill's and Marx's day.

Capitalism in each of the Western nations in the 1970s is experiencing economic difficulties of near-crisis proportions. Of these no end is in sight. Keynesian remedies, successful for three decades from the 1930s, have now evidently failed to cope with the underlying contradiction. The most obvious symptom of this failure is the prevalence, simultaneously, of high rates both of inflation and of unemployment—two things which used to be thought alternatives. For wage-earners, the erosion of the value of money earnings along with insecurity of employment is a serious matter. It has already led to increased working-class militancy in various forms: in some countries, increased political activity and strength of communist and socialist parties; in others, increased participation in trade union and industrial activity. The trade unions will be increasingly impelled not just to concern themselves with labour's share of the national income but to recognize the structural incompetence of managed capitalism. It cannot be said that trade union leaders generally have yet seen this, but they are being increasingly hard-pressed by shop steward activity and unofficial strike action. It is to be expected that working-class participation in political and industrial action will increase, and will be increasingly class-conscious. The probability is that industrial action, of which there is a lot already, will be seen to be fundamentally *political*, and so, whether it takes the form of participation in the formal political process or not, will amount to increased political participation.

So we have three weak points in the vicious circle—the increasing awareness of the costs of economic growth, the increasing awareness of the costs of political apathy, the increasing doubts about the ability of corporate capitalism to meet consumer expectations while reproducing inequality. And each of these may be said to be contributing, in ways we have seen, to the possible attainment of the prerequisite conditions for participatory democracy: together, they conduce to a decline in consumer consciousness, a reduction of class inequality, and an increase in present political participation. The prospects for a more democratic society are thus not entirely bleak. The move towards it will both require and encourage an increasing measure of participation. And this now seems to be within the realm of the possible.

Before leaving this discussion of the possibility of moving to a participatory democracy, I must emphasize that I have been looking only for possible, even barely possible, ways ahead. I have not attempted to assess whether the chances of winning through are better or worse than 50/50. And when one thinks of the forces opposed to such a change, one might hesitate to put the chances as high as 50/50. One need only think of the power of multi-national corporations; of the probability of the increasing penetration into home affairs of secret intelligence agencies such as the American C.I.A., which have been allowed or required by their governments to include in 'intelligence' such activities as organizing invasions of some smaller countries and assisting in the overthrow of disliked governments of others; and of the increasing use of political terrorism by outraged minorities of right and left, with the excuse they give governments of moving into the practices of the police state, and even getting a large measure of popular support for the police state. Against such forces can only be put the fact that liberal-democratic governments are reluctant to use open force on a large scale, except for very short periods, against any widely supported popular movements at home: understandably so, for by the time a government feels the need to do this it may well be unable to count on the army and the police.

At a less immediately alarming level there are other factors which may prevent the requisite reduction of class inequality. The advanced Western economies may slow down to a stationary condition (where there is no economic growth because no incentive to new capital formation) before popular pressures have done much to get the present class inequalities reduced: this would make further reduction more difficult. And the maintenance of even the present Western levels of affluence would be impossible if some of the underdeveloped nations were able, by nuclear blackmail or otherwise, to impose a redistribution of income between the rich and poor nations. Such a global redistribution would make still more difficult any significant reduction of class inequality within the affluent nations.[8]

I do not know of enough empirical evidence to enable one to judge the relative strength of the forces in our present society making for, and those making against, a move to a more participatory democracy. So my exploration of possible forces making for it is not to be taken as a prophecy, but only as a glimpse of possibilities.

MODELS OF PARTICIPATORY DEMOCRACY

Let me turn finally to the question of how a participatory democracy might be run if we did achieve the prerequisites. How participatory could it be, given that at any level beyond the neighbourhood it would have to be an indirect or representative system rather than face-to-face direct democracy?

(i) *Model 4A: an abstract first approximation*

If one looks at the question first in general terms, setting aside for the present both the weight of tradition and the actual circumstances that might prevail in any country when the prerequisites had been sufficiently met, the simplest model that could properly be called a participatory democracy would be a pyramidal system with direct democracy at the base and delegate democracy at every level above that. Thus one would start with direct democracy at the neighbourhood or factory level—actual face-to-face discussion and decision by consensus or majority, and election of delegates who would make up a council at the next more inclusive level, say a city borough or ward or a township. The delegates would have to be sufficiently instructed by and accountable to those who elected them to make decisions at the council level reasonably democratic. So it would go on up to the top level, which would be a national council for matters of national concern, and local and regional

[8] Cf. Robert L. Heilbroner: *An Inquiry into the Human Prospect*, 2nd edn., New York, 1975, especially ch. 3, where it is argued that, for reasons such as these, the Western nations are unlikely to be able to keep up even their present degree of liberal democracy.

councils for matters of less than national concern. At whatever level beyond the smallest primary one the final decisions on different matters were made, the issues would certainly have to be formulated by a committee of the council. Thus at whatever level the reference up stopped, it would stop in effect with a small committee of that level's council. This may seem a far cry from democratic control. But I think it is the best we can do. What is needed, at every stage, to make the system democratic, is that the decision-makers and issue-formulators elected from below be held responsible to those below by being subject to re-election or even recall.

Now such a system, no matter how clearly responsibilities are set out on paper, even if the paper is a formal national constitution, is no guarantee of effective democratic participation or control: the Soviet Union's 'democratic centralism', which was just such a scheme, cannot be said to have provided the democratic control that had been intended. The question is whether such failure is inherent in the nature of a pyramidal councils system. I think it is not. I suggest that we can identify the sets of circumstances in which the system won't work as intended, that is, won't provide adequate responsibility to those below, won't be actively democratic. Three such sets of circumstances are evident.

(1) A pyramidal system will not provide real responsibility of the government to all the levels below in an immediately post-revolutionary situation; at least it will not do so if the threat of counter-revolution, with or without foreign intervention, is present. For in that case, democratic control, with all its delays, has to give way to central authority. That was the lesson of the immediate aftermath of the 1917 Bolshevik revolution. A further lesson, to be drawn from the subsequent Soviet experience, is that, if a revolution bites off more than it can chew democratically, it will chew it undemocratically.

Now since we do not seem likely, in the Western liberal democracies, to try to move to full democracy by way of a Bolshevik revolution, this does not appear to be a difficulty for us. But we must notice that the threat of counter-revolution is present not only after a Bolshevik revolution but also after a parliamentary revolution, i.e. a constitutional, electoral, take-over of power by a party or popular front pledged to a radical reform leading to the replacement of capitalism. That this threat may be real, and be fatal to a constitutional revolutionary regime which tries to proceed democratically, is evident in the example of the counter-revolutionary overthrow of the Allende regime in Chile in 1973, after three years in office. We have to ask, therefore, whether the Chilean sequence could be repeated in any of the more advanced Western liberal-democracies. Could it happen in, say, Italy or France? If it could, the chances of participatory democracy in any such country would be slim.

There is no certainty that it could not happen there. We cannot rely on there being a longer habit of constitutionalism in Western Europe than in Latin America: indeed, in those European liberal democracies which are most likely to be in this situation in the forseeable future (e.g. Italy and France), the tradition of constitutionalism cannot be said to be much older or firmer than in Chile. We should, however, notice that Allende's popular front coalition was in control of only a part of the executive power (the presidency, but not the *contraloria*, which had power to rule on the legality of any executive action), and was in control of none of the legislative (including taxing) power. If a similar government elsewhere came into office with a stronger base it could proceed democratically without the same risk of being overthrown by counter-revolution.

(2) Another circumstance in which a responsible pyramidal councils system would not work would be a reappearance of an underlying class division and opposition. For, as we have seen, such division requires that the political system, in order to hold the society together, be able to perform the function of continual compromise between class interests, and that function makes it impossible to have clear and strong lines of responsibility from the upper elected levels downwards.

But this also is not as great a problem for us as it might seem. For if my earlier analysis is right, we shall not have reached the possibility of installing such a responsible system until we have greatly reduced the present social and economic inequalities.

It is true that this will be possible only in the measure that the capital/labour relation that prevails in our society has been fundamentally changed, for capitalist relations produce and reproduce opposed classes. No amount of welfare-state redistribution of income will by itself change that relation. Nor will any amount of workers' participation or workers' control at the shop-floor level or the plant level: that is a promising breakthrough point, but it will not do the whole job. A fully democratic society requires democratic political control over the uses to which the amassed capital and the remaining natural resources of the society are put. It probably does not matter whether this takes the form of social ownership of all capital, or a social control of it so thorough as to be virtually the same as ownership. But more welfare-state redistribution of the national income is not enough: no matter how much it might reduce class inequalities of income it would not touch class inequalities of power.

(3) A third circumstance in which the pyramidal council system would not work is, of course, if the people at the base were apathetic. Such a system could not have been reached except by a people who had thrown off their political apathy. But might not apathy grow again? There can be no guarantee that it would not. But at least the main factor which I have argued creates and sustains apathy in our present system would by hypothesis be absent or at least greatly modified—I mean the class structure which discourages the participation of those in the lower strata by rendering it relatively ineffective, and which more generally discourages participation by requiring such a blurring of issues that governments cannot be held seriously responsible to the electorate.

To sum up the discussion so far of the prospects of a pyramidal councils system as a model of participatory democracy, we may say that in the measure that the prerequisite conditions for transition to a participatory system had been achieved in any Western country, the most obvious impediments to a pyramidal councils scheme being genuinely democratic would not be present. A pyramidal system might work. Or other impediments might emerge to prevent it being fully democratic. It is not worth pursuing these, for this simple model is too unrealistic. It can be nothing but a first approximation to a workable model, for it was reached by deliberately setting aside what must now be brought back into consideration—the weight of tradition and the actual circumstances that are likely to prevail in any Western nation at the time when the transition became possible.

The most important factor here is the existence of political parties. The simple model has no place for them. It envisages a no-party or one-party system. This was appropriate enough when such a model was put forward in the revolutionary circumstances of mid-seventeenth-century England and early twentieth-century Russia. But it is not appropriate for late twentieth-century Western nations, for it seems unlikely that any of them will move to the threshold of participatory democracy by way of a one-party revolutionary take-over. It is much more likely that any such move will be made under the leadership of a popular front or a coalition of social-democratic and socialist parties. Those parties will not wither away, at least not for some years. Unless all of them but one are put down by force, several will still be around. The real question then is, whether there is some way of combining a pyramidal council structure with a competitive party system.

(ii) *Model 4B: a second approximation*

The combination of a pyramidal direct/indirect democratic machinery with a continuing party system seems essential. Nothing but a pyramidal system will incorporate any direct democracy into a nation-wide structure of government, and some significant amount of direct democracy is required for anything that can be called participatory democracy. At the same time, competitive political parties must be assumed to be in existence, parties whose claims cannot, consistently with anything that could be called a liberal democracy, be overridden.

Not only is the combination of pyramid and parties probably unavoidable: it may be positively desirable. For even in a non-class-divided society there would still be issues around which parties might form, or even might be needed to allow issues to be effectively proposed and debated: issues such as the over-all allocation of resources, environmental and urban planning, population and immigration policies, foreign policy, military policy.[9] Now supposing that a competitive party system were either unavoidable, or actually desirable, in a non-exploitive, non-class-divided society, could it be combined with any kind of pyramidal direct/indirect democracy?

I think it could. For the main functions which the competitive party system has had to perform, and has performed, in class-divided societies up to now, i.e. the blurring of class opposition and the continual arranging of compromises or apparent compromises between the demands of opposed classes, would no longer be required. And those are the features of the competitive party system which have made it up to now incompatible with any effective participatory democracy. With that function no longer required, the incompatibility disappears.

There are, in abstract theory, two possibilities of combining a pyramidal organization with competing parties. One, much the more difficult, and so unlikely as to deserve no attention here, would be to replace the existing Western parliamentary or congressional/presidential structure of government by a soviet-type structure, (which is conceivable even with two or more parties). The other, much less difficult, would be to keep the existing structure of government, and rely on the parties themselves to operate by pyramidal participation. It is true, as I said earlier, that all the many attempts made by democratic reform movements and parties to make their leaders, when they became the government, responsible to the rank-and-file,

[9] It is worth noticing that in Czechoslovakia, in the spring and summer of 1968 just before the overthrow of the reformist Communist Dubček regime by the military intervention of the U.S.S.R., one of the widely canvassed proposals for enhancing the democratic quality of the political system was the introduction of a competitive party system, and that this had substantial public support and even some support within the ruling Communist Party. In a July public opinion poll 25 per cent of the C.P. members polled, and 58 per cent of non-party persons polled, wanted one or more new parties; in an August poll, in which the question was put ambiguously, the figures were 16 per cent and 35 per cent. (H. Gordon Skilling: *Czechoslovakia's Interrupted Revolution*, Princeton University Press, 1976, pp. 550–1, 356–72.)

have failed. But the reason for those failures would no longer exist in the circumstances we are considering, or at least would not exist to anything like the same degree. The reason for those failures was that strict responsibility of the party leadership to the membership does not allow the room for manœuvre and compromise which a government in a class-divided society must have in order to carry out its necessary function of mediating between opposed class interests in the whole society. No doubt, even in a non-class-divided society, there would still have to be some room for compromise. But the amount of room needed for compromise on the sort of issues that might then divide parties would not be of the same order of magnitude as the amount now required, and the element of deception or concealment required to carry on the continual blurring of class lines would not be present.

It thus appears that there is a real possibility of genuinely participatory parties, and that they could operate through a parliamentary or congressional structure to provide a substantial measure of participatory democracy. This I think is as far as it is now feasible to go by way of a blueprint.

PARTICIPATORY DEMOCRACY AS LIBERAL DEMOCRACY?

One question remains: can this model of participatory democracy be called a model of liberal democracy? I think it can. It is clearly not dictatorial or totalitarian. The guarantee of this is not the existence of alternative parties, for it is conceivable that after some decades they might wither away, in conditions of greater plenty and widespread opportunity for citizen participation other than through political parties. In that case we should have moved to Model 4A. The guarantee is rather in the presumption that no version of Model 4 could come into existence or remain in existence without a strong and widespread sense of the value of that liberal-democratic ethical principle which was the heart of Model 2—the equal right of every man and woman to the full development and use of his or her capabilities. And of course the very possibility of Model 4 requires also, as argued in the second section of this chapter, a downgrading or abandonment of market assumptions about the nature of man and society, a departure from the image of man as maximizing consumer, and a great reduction of the present economic and social inequality. Those changes would make possible a restoration, even a realization, of the central ethical principle of Model 2; and they would not, for the reason given earlier,[10] logically deny to a Model 4 the description 'liberal'. As long as there remained a strong sense of the high value of the equal right of self-development, Model 4 would be in the best tradition of liberal democracy.

[10] At the end of ch. I, pp. 21–2.

Further Reading

Those who want to get further into a subject like this, which is both analytical and historical, will generally find it more rewarding to go first to some of the works of the leading original writers rather than relying on even the best secondary accounts of them, especially when, as is sometimes the case, the former are shorter than the latter.

To appreciate the enormously confident style of the early nineteenth-century theorists of liberal democracy one could not do better than to look at James Mill's famous article 'Government' (written first for a supplement to the fifth edition of the *Encyclopaedia Britannica* in 1820 and reprinted many times, usually as *An Essay on Government*), or a few pages of Bentham—either the brief chapters of his *Principles of the Civil Code* cited above in ch. II, nn. 2, 7–12, and 15–18, or the first few chapters of his *Introduction to the Principles of Morals and Legislation*.

The classic statement of Model 2A is John Stuart Mill's *Considerations on Representative Government*. The most elegant short presentation of Model 2B is A. D. Lindsay's *The Essentials of Democracy*. There is a useful account of some further 2B theorists in ch. I of Dennis F. Thompson's *The Democratic Citizen*, London, Cambridge University Press, 1970.

The leading expositions of Model 3 are the works listed in nn. 1 and 2 of ch. IV: the best are still Schumpeter's ch. 22 and Dahl's short *Preface to Democratic Theory*. The leading critiques of Model 3 are the works listed in n. 11 of ch. IV: each of the three collections of essays cited there affords an excellent statement of the case against Model 3. My short *The Real World of Democracy*, and Essay 10 in my *Democratic Theory: Essays in Retrieval*, put Model 3 in an unflattering global perspective.

Realistic works on participatory democracy are scarce. Its advocates incline simply to celebrate direct democracy, often as a way towards an ideal anarchistic society (for example in many of the essays in C. George Benello and Dimitrios Roussopoulos (eds.): *The Case for Participatory Democracy: Some Prospects for a Radical Society*, New York, Grossman, 1971). But there are useful treatments in Carole Pateman's *Participation and Democratic Theory* and in the *Nomos* volume *Participation in Politics* cited in n. 3 to ch. V. An earlier volume, also entitled *Participation in Politics*, edited by Geraint Parry (Manchester University Press, 1972), has interesting essays on the possibility and desirability of more participation, on the place of participation in Marxian theory, and on the record in some Western and Communist and Third World countries.

Index